I KNOW THE AN

SIMON KHACHATRYAN

COPYRIGHT © 2016 by Simon Khachatryan

All rights reserved.

This book is based on the precious works of **Hugh Auchincloss Brown**, **Immanuel Velikovsky**, **Zecharia Sitchin, Hans Bellamy** and **Karl Brugger.** I also include major events from the Old Testament.

I provide in my book scientific arguments that go against widely accepted theories of the origin of continents, oceans, mountains, lakes and volcanoes. The reader of this book will find a revolutionary approach to the problems of appearance of our universe, origins of every celestial body and their eventual fate.

With detailed descriptions of the purpose of all the megalithic structures on Earth, the book makes an undoubted breakthrough in the revelation of all the major enigmas of antiquity.

I invite you to join me on this journey to making your own judgment.

Thanks to the development of satellite imaging, the discoveries of explorers, and the researchers whose precious works led to the writing of this book.

Table of contents

Introduction 1

Academic Science 3

Human Origin 4

The Earth 8

Reversal 14

"Imagination is more important than knowledge." 22

Lakes 29

Ice Age 32

Strange Geography 34

"Vampire Ocean" 40

Prehistoric Andes 45

The First Great Catastrophe 51

The Water 54

The Scars of the Earth 56

Frozen in Time 72

Earthquakes 77

Volcanoes 78

The Moon 94

Mars 104

Satellites 109

Planets and Stars 111

Who Are We? 122

Prehistoric Giants...137

The Pyramids 141

Dolmens 150

Boulders 156

Ancient Structures 160

Offensive Truth...173

Atlantis 182

Recent Polar Shift 189

The Second Great Catastrophe...191

The Ark 195

Epilogue...208

Chronological Table of Major Events on Earth 217

Introduction

From our early childhood we were inspired by myths and stories about ancient adventurers and hidden treasures, extinct mammoths and dinosaurs, pyramids and ancient cities, distant stars and mysterious galaxies. During the first half of our lives, we believed that our parents were hiding many answers from us and that one day when we grew up all our questions would be answered. Time passed by, and we grew up and became more experienced and more knowledgeable in many aspects of our lives. When we hit our (so unwelcomed) 40s, we realized that we are on the top of our "life pyramid" and soon, slowly but steadily, we were going to go down. Gradually the roles are changing and the next generations are asking the questions we were asking our parents. By trying to answer the question: "Who are we and for what purpose are we here?" we realize that we have no clue about our source of origin, except the one described in the Old Testament. After communicating with older and presumably wiser generations we realize that all their wisdom does not go beyond human relationship, politics and philosophy, while the main question of our appearance on this planet remains unanswered.

In search for answers, we realize that except for the common desire of happiness for ourselves, the meaning of our lives gradually transforms into the process of providing better life conditions for our direct descendants (our children and grandchildren), and the fate of generations after them strangely becomes less of our concern. Happiness is a process of growth, whether it is physical, mental, financial or any other type of growth. After reaching its apogee, any decline of the former growth leads to an eventual loss of interest in life. Eventual admission of our helplessness before Her Majesty — the "Time" — and awareness of our eventual approach to the final destination gradually leads to our self-reassuring and revengeful defiance to the heavens: "I will pass away, but my descendants will live on." With this self-calming slogan millions and millions of people are passing away to nothingness.

If we look at all living things on Earth from a computerized point of view, we can describe ourselves as carriers of different programs. A newborn child is nothing more than a "brand new memory chip" that observes and collects information he receives from the first days of his life. The recording of this information occurs involuntarily, and we have no control over our craving for information. That is the reason why every parent desires to fill his offspring's mind with a mentality similar to his own; otherwise it will be filled with who knows what. Independent of race and nationality, a child born in a different society absorbs the language, traditions and understanding of that particular environment.

During our lives, by "downloading" this or that "program" we are forming our personalities and therefore our physical and intellectual characteristics can be described as:

We are what we eat (food wise), and we are what we absorb (information wise).

Being capable of weighing all the facts we collected through the existence of our civilization, we can explain the appearance of our universe and therefore reconsider our own appearance on this planet. All we have to do is to accept the arguments of our ancestors and try to look at the subjects they have described through their own points of view.

This book is based on the precious works of **Hugh Auchincloss Brown**, **Immanuel Velikovsky**, **Zecharia Sitchin** and **Karl Brugger.** I also include major events from the Old Testament.

I provide in my book scientific arguments that go against widely accepted theories of the origin of continents, oceans, mountains, lakes and volcanoes. The reader of this book will find a revolutionary approach to the problems of appearance of our universe, origins of every celestial body and their eventual fate.

With detailed descriptions of the purpose of all the megalithic structures on Earth, the book makes undoubted breakthrough in the revelation of all the major enigmas of antiquity.

I invite you to join me on this journey to making your own judgment.

Academic Science

Nowadays we hear more and more criticism about academic science and its bureaucratic apparatus. Whatever does not fit into its proven dogmas is being automatically dismissed. The way we live today is also thanks to the same academic science and technology. Of course it is far from perfect, and maybe there are many other and more advanced and cheaper technologies to be realized in our lives, but to "demolish" existing science without introducing something new means going back to the Stone Age period again.

Archaeologists constantly find artifacts, which cannot be explained and therefore cannot be linked to any existing scientific theory. If there is no explanation of their origin, they are stored in archives. But science was created by us humans. If we see an object we have never seen before, our brain begins to look for similar silhouettes in our memory to compare it with the image we see. If the search does not yield any results after a while, we try to convince ourselves that it was not real, and if we never see the object again we even begin to question its existence. The rest is a matter of time. Time cures.

Similarly with our mind, science also has its trump card — Her Majesty the "Time." For example, according to science the Earth and moon were created four and a half billion years ago. For those of us whose average lifespan is around 70 years, millions and billions of years means something unreachable and hardly imaginable. Problem solved. Nobody else wants to ask any questions.

Paleontologists constantly discover dinosaur fossils in different strata of the earth. Sedimentary rocks formed by the deposition of material on the earth's surface take hundreds of years to harden. Considering each stratum as a period of time, scientists came to the conclusion that dinosaurs lived on Earth "from 231.4 to 66 million years ago." But that principle has the right to exist only in one case: millions of years of absolute stable environmental conditions on Earth.

Comparing our lifespan to the age of our planet, we consider the earth as some kind of constant. But should we? Has the earth always rotated in the same direction? Were there any shifts of poles on Earth? Was the sea level always the same? Was the moon always there? And so on and so forth. We can ask hundreds of questions and not get any reasonable answers.

According to the "Old Testament" and many ancient manuscripts, there were many events and mythical characters that modern science can neither prove nor disprove. That is the major non-compliance between science and the religion we created. There is a huge gap between them, and it is understandable why that is so because to get a scientific explanation of an event is much harder than just to witness it. That is why science is way behind religion, and our only hope for the future is to shorten the gap between them. Every generation has a hard time perceiving such double, sometimes completely opposing foundations: religious and scientific. And in this book I will try to bridge that gap as much as I can.

Human Origin

Unlike animals, we can work, write and read, which allows us to pass information from generation to generation without ever seeing each other. That is why we constantly look for ancient sources in the hope of finding answers of our origin. However, because the information we get from ancient sources differs from our reality, we accept it as a myth or a tale and nothing else. There are two primary versions of human origin: evolutionary and biblical.

Evolutionary

Perhaps the fact that every generation, for some strange reason, considers itself much smarter and more advanced than older generations, inspired Charles Darwin to create his famous theory. But even the most atheistic person, at the back of his mind agrees with God's creationism rather than evolving from monkey. With 96 percent of similar DNA cells between us and apes, we have around 40 million differences which could not develop just by themselves with the passage of time. Even if we consider the so-called Homo habilis as our earliest known ancestor, who appeared 2.3 million years ago, then during all those years we were supposed to develop around seventeen differences a year. But it does not happen. That means there was either a sudden mutation or an artificial change in our genes 2.3 million years ago (if those numbers are correct).

With our pseudoscience we are going nowhere, but to dismiss Darwin's theory completely would be wrong because indeed so many animal species on earth have numerous similarities and to call it just accidental coincidence would not be intelligent. And therefore Darwin's evolution of species is absolutely right only in one case: major changes of our planet's environment such as the changing of axis, direction of rotation, speed of rotation, distance to the sun and so on. Changes in climate or magnetic field could trigger drastic mutations of species because genes cannot just mutate without any cause. Any object resting in peace will never move unless some force is applied.

We can surely accept the theory of evolution, but what does not make sense in Darwin's theory is the emergence of humans. We can definitely assume that there were sudden changes of climate, the magnetic field, et cetera, but none of the above mentioned factors could trigger the appearance of intellect. The disappearance of food on the trees could not lead monkeys to cultivate the land even in millions of years. There are so many animal species that have just vanished from the surface of the earth because of lack of food or water, and strangely none of them had ever tried to cultivate anything. Considering the above mentioned, we can make the assertion that **mutation of any living thing on Earth is the result of major cataclysmic changes in the environment.**

Biblical

"God created man in his own image, in the image of God created he him; male and female created he them."

(Gen. 1: 27)

Many scholars consider the biblical "forbidden fruit" as forbidden sex, because right after eating it Adam and Eve attained a sense of shame from being naked. Basically God created humans with reproduction abilities and got mad that people discovered it. What nonsense!

If "God created us in his own image," it would be logical to assume that he created us with his own way of thinking also, isn't it? And if today we were able to create living robots to help us in our everyday life, we would create just helpers rather than creatures smarter than us. Just as with the appearance of computers in our lives, we have anxiety about eventually being destroyed by them. It is just simple fear that one day they will stop obeying us. Gods could have exact same concerns about humans.

"You may surely eat of every tree of the garden, but of the tree of the knowledge of good and evil you shall not eat, for in the day that you eat of it you shall surely die." (Gen. 2:16)

In 1976 Karl Brugger published his book *The Chronicle of Akakor*. The book describes the life and struggle of people of Ugha Mongulala, the natives of South America with a 12,500-year-old history. I came across very curious details in the book. When high priests were gathering to fast in order to predict the future, *"They ate only three kinds of fruit and small corn cakes" (page-85)*. So why don't we perceive the meaning of "forbidden fruit" literally? The absence of that particular "forbidden tree" at the present time makes us look for some hidden meanings. I am writing this book in the hope that my language will be attainable for anyone. Then why do we think that ancient authors wrote the Old Testament only for the chosen ones?

There is also reviewed biblical version where of course we have to mention the name of Zecharia Sitchin (1920 – 2010) Author of *The 12th Planet*, (1976). According to his works there was a race of extraterrestrials (Anunnaki) from planet Nibiru, who came to Earth in search of gold, which they needed desperately in order to save their own planet's atmosphere. After years of hard labour in Earth's gold mines, the Anunnaki decided to create primitive workers to help them in their multiple tasks on our planet.

Despite the outrage of many scholars, there is nothing irrational about this theory. Nowadays our civilization is pretty advanced in manipulating genes, and we actually do. But we do not create genetically incubated creatures just for fun or in order to play the role of gods to them. "The apple does not fall far from the tree." We are supposed to have the same way of thinking as our creator, and therefore it makes sense to create something useful for us, isn't it? We manufacture machines and computers to help us in the fulfillment of different tasks. The need of physical help could push God to conduct genetic experiments. According to Z. Sitchin, before the appearance of the first successful human (Adam), god(s) had worked and created numerous other forms of life who coexisted with first humans. The extinction of defective and odd looking forms of life led to the appearance of mythology.

Have you ever tried to imagine bizarre looking animals by combining different parts of two or more animals? It is too childish and complete nonsense, isn't it? Then why would our ancestors keep themselves busy with such thoughts? Maybe they have really seen them? If we draw a parallel between us and God, their need in various biological robots is quite explicable. Centaurs, pegasus, mermaids, winged people, little people, giant people, etc. All of them could serve in different areas of God's life. As creatures of the land that always had a desire to fly, swim and dive, we also create our computerized helpers and along with mini-excavators we created excavators of immense sizes. Curiously enough with each archaeological discovery of odd-looking human remains, we always convince ourselves that we are the originals and the rest of them are just abnormal.

The other thing that confirms Mr. Sitchin's hypothesis is the human reproductive ability. With an "all-season" mating ability, the human population is constantly growing. Many scholars claim it as the result of abundance of food, but reality shows the opposite. Economically advanced countries have lower birth rates than countries that are struggling. We have exactly the same protective functions as anything else on our planet. The tree growing on a cliff with a lack of nourishment has longer and stronger roots than the tree growing on the rich soil. So somebody programmed us to breed like rabbits and set our system of self-preservation at higher than average levels. Just imagine if the production of automobiles were seasonal, it would never fulfill our needs.

The Earth

   Hugh Auchincloss Brown (1879 – 1975), is the author of *Cataclysms of the Earth* (1967). According to his theory, every 3,000-7,000 years the Earth experiences recurrent careening. The main cause of it is the constantly increasing weight of polar ice. After reaching its maturity, thanks to the centrifugal force, enormous masses of ice make the Earth career on its side, shifting geographic poles to the equator in a matter of days or maybe even hours. The proof of his theory are the radial striations on Earth, which are basically huge dents made by the enormous weight of former polar ice caps. According to his theory, there are four most recent poles on Earth: Caspian Sea Depression, Sudan Basin, Hudson Bay Basin and the present North Pole. The fact that Earth experienced many polar shifts in the past is obvious. Layers of salt in our strata prove that land was periodically under seawater.

   Nowadays, our civilization is developed enough to explore outer space. Meanwhile, how many of us know how mountains are formed? Of course what comes to mind is a volcano, which can create one or several mountain peaks, but what about mountain ranges?

   The scientific explanation is: *"Colliding of tectonic plates pushes the land masses upward, creating mountain ranges; a process which takes millions of years."* But why can no one explain the source of such force capable to push the land masses against each other? Acceptance of this theory automatically makes our planet 4.5 billion years old. But is that so?

Hugh Brown explains that differently:

"after 90 degree careening... kinetic energy which has developed in the continental land masses because of their weights and velocities, collides with the combined energy of the newly generated bulges of the earth and of the ice caps.

The result of these collisions of forces is that the energy of the moving continents is absorbed by the crushing, elevating, and wrinkling of large land areas..." (page 102)

It is so unfortunate we did not study Hugh Brown's theory at school instead of some invisible "battle of tectonic plates". People have the same instincts as animals, and one of the first lessons of childhood is to never run if you see a dog or any other predator. Motionless creatures do not interest anybody, and we live by the same rules. For us, for instance, watching running cheetahs is more attractive than watching the crawling turtle. Analogically, it seems that somebody created a theory of very-slow moving tectonic plates in order to put our conscious mind to sleep.

Half a century ago this unrecognized scientist proved the theory of shifting poles (Hugh Brown's book was published in 1967). It is strange that after such a long period of time we still do not pay serious attention to his theory. Maybe in order to achieve scientific recognition, one is to follow the same path as academic science? (He was not a scientist at all). In other words, in order to become an army general, naturally one would have to start as a soldier. Nobody likes smart outsiders, and I am sure not everyone will like my book also.

In hundreds of years, snow has been constantly accumulating on geographic poles and after an x-amount of time, recurrent reversal shifts the location of the poles to warmer latitudes. During the polar shifts, inertia of heavy poles interferes with the complex movement of the reversing Earth, where, thanks to centrifugal force, the heaviest part of the globe slides towards the equator of the planet. Gradually, after the shifting of geographical poles, the former equatorial bulge begins to resolve back into the mass of the Earth while the newly developing bulge is trying to make its way through the massive and heavy body of polar ice located on the equator.

The massiveness of the ice cap located on the top of the newly developing equatorial bulge makes an enormous dent in it, until it loses its weight because of a **rapid** melt down. After complete disappearance, the former ice cap leaves its round pattern pressed into the piled up mountains of rock. And we call those "battlefields" between the crushing powers of ice and rock mountain ranges. Our present equatorial bulge is 42.72 km or 21.36 km on each side. Just imagine the equatorial bulge trying to grow from a couple of kilometres to 21 km high. The force is enormous.

(Pic.1) Mountains of Zhangye Danxia. China

In that time, sediment deposits create horizontal layers. Mountains with tilted sedimentary lines indicate that formerly horizontal lines were pushed and piled up by an enormously powerful force. One of the best and most spectacular proofs of that is the multi-coloured mountains of Zhangye Danxia in China, located 1800 meters above present sea level *(Pic.1)*. It seems that even nature gives us an important hint to unravel the origin of mountains. All the tilted lines indicate their pushed and piled up nature where immense power of centrifugal force pushed enormous masses of polar ice towards a newly developing equatorial bulge. The highest mountains on Earth are just the result of the collision of a newly developed equatorial bulge with the ice masses of the previous pole.

According to Hugh Brown's theory, the weight of polar ice leads to recurrent reversals of the Earth. What I would add to his theory is that the fate of future reversal and the time between reversals depends on the location of present geographical poles. Poles located in water or on dry land cannot have the same periodicity. Let us consider each of these scenarios with respect to the equatorial bulge.

1.   **Both poles are located on land.** For centuries, along with its constantly increasing weight and being widespread around poles, enormous masses of snow also grow in height, and all the "fuel" for its accumulation comes from equatorial regions of the globe. In cases where poles are located on dry land and equatorial regions are covered by oceans, there will be a massive process of the pumping of global waters from the equator to the poles. Usually evaporated waters of the oceans fall back to the earth as precipitation creating a continuous cycle. Without water exits, accumulated snow from the polar regions stays there forever, thereby breaking the chain of water cycles. Taking into account that every spinning planet has an equatorial bulge, the pumping of equatorial waters to the poles will lead to gradual elevation of the poles. The relatively uniform shape of the globe will gradually transform into an irregularly-shaped sphere with much heavier poles (a boiled egg does not spin on its sharp point). Attempts of the centrifugal force to pull the heaviest objects of the globe towards the equator will succeed when the balance between the equatorial bulge and geographical poles reaches its point of instability, and the globe will careen on its side.

2.   **One of the poles is located on water while the other one is on dry land.** A geographical pole located on dry land will accumulate snow and grow in size, weight and altitude. In contrast, snow accumulation of the pole located in the ocean will never reach its maximum value. After reaching their breaking points, massive icebergs will eventually separate from the main body of polar ice, and under the influence of the centrifugal force they will be carried towards the equator. After centuries of such

an uneven distribution of weight on the globe, the balance of the spinning planet will eventually reach its point of instability. The globe will start to wobble and the pulling momentum of the centrifugal force will eventually shift the heavy pole (located on dry land) towards the equator.

3. Both poles located in the ocean. This scenario can be considered as the safest and having the longest period of stability, where the weights of both poles are released by periodically separating from the main polar cap icebergs. Never reaching their maximum value, the weights of both polar caps will counterbalance each other.

Locations of present poles can be considered as the combination of scenarios two and three. Even though one of the poles is located in the ocean and the second one on dry land, luckily for our civilization the size of the Antarctic continent does not allow accumulated ice to spread any farther, and constantly separating icebergs keep the area of polar snow considerably changeless. Here we have to introduce two additional sub-groups:

a) Factor of size of water body and size of continent under the pole, which plays a significant role in formation of polar ice.

b) Closer proximity of poles to the shores of continent lead to higher risk of the Ice Age effect.

According to the three scenarios mentioned above, the most severe and continuous Ice Age takes place when both poles are located on dry land. When poles are located in different environments, one in ocean and one on land, the planet will experience a less severe Ice Age, at least on one hemisphere. The period much more continuous and more suitable for living is the scenario with both poles located in the ocean. The present climate of our planet can be considered moderate in the Northern hemisphere with a small Ice Age effect in the Southern Hemisphere.

All the former poles which happened to be located in water are untraceable because they just melted down without any landmark, but the former poles located completely on dry land without access to the sea indicate the existence of severe Ice Ages in the past.

During recurrent reversals, previous equatorial bulges were dissolving back into the Earth's surface and newly-created ones, on the other hand, facing enormous masses of ice above, were protruding up wherever they could. And because of the round shape of polar ice, the mountains which were appearing from the sides also were taking the shape of an arc. All the arc-shaped mountain ranges on our planet are silent proofs of that process.

In his book, Hugh Brown was indicating the three most recent former poles on Earth: Hudson Bay Basin, Caspian Sea Depression and Sudan Basin. Some of them are doubtful but we will go over them later in this book. Today, thanks to the development of satellite imaging, besides the above mentioned former poles, it is much easier to spot four other former geographical poles. All we have to do is just look for rounded shapes of mountain ranges.

1.   **Amazon Basin** - one of the most obvious former poles, with almost perfect semi-round Andean mountains wrapped around it from the west.

2.   If we place the centre of a former pole around the city of **Smolensk** in Russia, we can measure almost the same distance to the north — to "Jumping Tiger's Back" (Scandinavian mountains), Dinaric Alps and Balkan Mountains on the west, Anatolian Plateau and Caucasus on south and Ural Mountains on the east. Of course the shape of the mountains is far from a perfect circle but the cause of it will be discussed later. Taking into account that the pole was located completely on dry land, which in contrary to the Antarctica of today, where huge ice lumps are constantly separating and floating away from the continent and thereby relieving the ice cap weight, we can easily assert that Europe at that time was experiencing an Ice Age.

3.   **Siberia**, Russia. The existence of such a massive mountain range as the Himalayas proves the existence of a pole in the past somewhere between Novaya Zemlya and Lake Baikal, Ural Mountains on the west and Verkhoyansk Range on the east. According to the size and height of the Himalayas Mountains, we can assert that a former polar cap had quite an impressive size and the Ice Age was very harsh.

4.   **Antarctica**. Scanned images of iceless Transantarctic Mountains in Antarctica show an obvious arch shape again, but that mountain range does not fit into its logical place. The arch-shaped Transantarctic Mountains are interrupting. But in order for the polar cap to make an imprint into a newly created equatorial bulge and push the land masses up, the previous polar ice has to be located on land and not in the ocean; otherwise there would be an opposite effect, and the newly created bulge would push the floating polar ice away. In order to push something heavy, we need a good foothold, don't we? The conclusion we can make here is that at the time when a newly developing bulge was creating the Transantarctic Mountains, the Antarctic continent was way bigger than it is now. So at some point in the past, the northern part of the Transantarctic Mountains was dry land. And it couldn't be just an island or archipelago. If a newly formed equatorial bulge had pushed and piled up the land upwards at a staggering 4,500 metre altitude, the former polar cap had to have a very massive weight, and that would be possible only if it were located on a continent. **But it is missing there.**

Reversal

In 1985 during the unloading of the orbital station, Salyut-7, Soviet cosmonaut, Vladimir Dzhanibekov, noticed some strange behaviour of a spinning cargo nut in space. Rotating in one direction, it was suddenly flipping upside down and was continuing to spin in the same direction of rotation even though it was now moving backwards. After repeating this experiment numerous times, the cosmonaut got the same results. That strange behaviour of spinning objects in a weightless environment was named the Dzhanibekov Effect.

Nowadays there are many Dzhanibekov Effect videos on the Internet, and if we look closely, we can notice that one of the poles stays in the same spot while the other pole makes a 180-degree jump behind the first one. Slow motion of an object experiencing the Dzhanibekov Effect shows that the direction of rotation in fact stays consistent. But what does change here is the position of the object towards the axis of rotation. Only one pole of the object begins to wobble around its axis; meanwhile some invisible force holds the other pole in the same spot.

It would be logical to suggest that the Dzhanibekov Effect is universal, and along with small objects every celestial body experiences the same effect, but there is one major exception. Here it is necessary to mention that all the objects experiencing the Dzhanibekov Effect in space have irregular shapes. The same cargo nut of Salyut-7 had two flat handles on it for its effortless usage.

At the same time many experiments show that external force applied on a spinning rotor does not affect its steady speed and has no influence on the direction of its rotation whatsoever. In contrary, any round object spinning in space acquires a surprisingly stable axis position, and an applied external force can push the object away but can never change its direction of rotation or inclination of its axis. But this effect applies on a uniform structure of rotor only. Any irregular shaped object thrown into space makes chaotic moves. That effect was used by ancient warriors when throwing two stones tied together with a rope. Such a weapon thrown with high speed acquires an unpredictable trajectory.

The theory proposed by Hugh Brown describes the same effect where the weight of accumulated polar ice does change the planet's centre of gravity. One of the main differences between spherical objects spinning in space and celestial bodies is that objects **do not** change their centres of gravity, meanwhile due to the accumulation of snow, all celestial bodies **do**. Experiments with a gyroscope spinning on a flat surface show that the lower pole stays in the same spot despite the turbulences of the upper pole especially during its slowing down period. In earthly conditions, after slowing down, the gyroscope eventually lands on the surface. On the contrary, in a gravitational environment where the freedom of a spinning gyroscope on a flat surface is limited by the surface it is situated on, open space allows the gyroscope to flip freely in every direction.

The seemingly impossible disturbance of a spinning rotor of an experimental gyroscope in space can be achieved by adding some adhesive weight to its poles. Any changes in the object's centre of gravity will have an impact on its rotation. Our planet is not one solid piece of rock. It consists of high rise mountains and lowland areas filled with water that is less dense than rock. Our planet's present axis of rotation is in conformity with its differently-elevated mountains and lowlands. The accumulation of heavy weight on the poles will gradually change the balance of the globe. The suggestion that the redistribution of weight on any celestial body cannot affect its axis of rotation would be absolutely wrong. Unfortunately, the unwillingness to recognize Hugh Brown's theory continues to get us nowhere.

Ancient witnesses

The phenomena of reversals were experienced by our ancestors in the past, and fortunately for us, many ancient texts prove the occurrences of such events. Many of them were collected by Immanuel Velikovsky in his book *Worlds in Collision* (1950). This unique collection of events confirms the occurrence of reversals in the past.

Papyrus Ipuwer says that "the Earth turned over like a potter's wheel" and "the Earth is upside down." (Page 127)

The Magical Papyrus Harris: "the south becomes north, and the Earth turns over" (Page 118)

Sophocle (Atreus): "Zeus ... changed the course of the sun, causing it to rise in the east and not in the west." (Page 121)

Plato, The Statesman (Politicus): "At certain periods the universe has its present circular motion, and at other periods it revolves in the reverse direction. Of all the changes which take place in the heavens this reversal is the greatest and most complete." (Page 120)

And the sun stood still, and the moon stayed, until the people had avenged themselves upon their enemies. Is not this written in the book of Jasher? So the sun stood still in the midst of heaven, and hasted not to go down about a whole day. (Joshua 10:12-14).

Pomponius Mela, a Latin author of the first century, wrote: The Egyptians pride themselves on being the most ancient people in the world. In their authentic annals... one may read that since they have been in existence, the course of the stars has changed direction four times, and that the sun has set twice in the part of the sky where it rises today." (Page 118)

It is worth mentioning that Pomponius Mela was a Roman geographer, who presumably wrote his work around 43 AD. The earliest mention of ancient Egypt goes as far back as 3100 BC. The citation *"the course of the stars has changed direction four times"* brings the average recurrence of Earth's reversal at the approximate rate of every 750 years or so.

"In the Mexican Annals of Cuauhtitlan — the history of the empire of Culhuacan and Mexico, written in Nahua-Indian in the sixteenth century — it is related that during a cosmic catastrophe that occurred in the remote past, the night did not end for a long time.

The biblical narrative describes the sun as remaining in the sky for an additional day ("about a whole day"). The Midrashim, the books of ancient traditions not embodied in the Scriptures, relate that the sun and the moon stood still for thirty-six "itim", or eighteen hours, and thus from sunrise to sunset the day lasted about thirty hours." ("Worlds in Collision", page 63)

Depicted evidence of Earth's reversal has survived till our days in the astronomical ceiling panel of Senenmut in Egypt, where the direction of constellations is opposite to the present day direction. The ceiling decoration dates back to the 15th Century BC and this means that at the time of decoration our planet was spinning in the opposite direction. The unfinished condition of the Senenmut panel could indicate the meaninglessness of its completion due to radical changes in Earth's rotation.

Archaeological evidence of shifting poles can be found in many near-coastal sunken cities around the globe. Two of them are the sunken cities of ancient Egypt — Canopus and Heracleion. At the time of their prosperity, the sea level was much lower than it is today and the only possibility for this to happen would be the shifting of the pole(s) from water to dry land. The location of the pole(s) on dry land leads to the massive accumulation of snow on it, in contrast to the pole(s) located in the ocean, where constantly breaking off icebergs are easing the weight of polar ice. Here we can certainly assert that at the time when these two cities were built (approximately 12th Century BC), the geographic poles were different from the present-day location, where at least one of the poles was located either on dry land or between two continents. Constantly accumulating snow on the pole(s) was gradually pumping water from the rest of the planet. Water that evaporated from around the globe was precipitating at the poles and because of the absence of outflow was staying there until the next polar shift, which would shift the poles to warmer latitudes.

In his book *Worlds in Collision*, Immanuel Velikovsky also pays attention to Exodus from a scientific point of view. Why would waters of the Red Sea retreat and then return as giant waves? If it is to suggest that in a matter of hours our globe experienced a polar shift, then the waters of the Red Sea could easily retreat and come back again. A commonly established period of Exodus is 14th Century BC.

Another incident took place during the reign of Hezekiah at the time of the siege of Jerusalem by the Assyrian king Sennacherib, who mysteriously lost his 185,000 army overnight by some strange force.

"Herodotus was told that myriads of mice descended upon the Assyrian camp and gnawed away the cords of their bows and other weapons; deprived of their arms, the troops fled in panic".

(Worlds in Collision, page 232)

As we know, mice are major transmitters of the bubonic plague, and there is some strange and very consistent pattern in our history:

- 22nd Century BC – intense, centuries-long droughts in Egypt, the Middle East, India and sudden cooling in China

- 14th Century (1446-1313 BC) – Exodus (retreat and following return of the waters of Red Sea)
- 7th Century (701 BC) – Destruction of Sennacherib's army (185,000 killed overnight)
- 1st Century BC – Birth of Christianity. Triggered by reversal temporary mass loss of memory could lead to necessity to regroup together again (hypothesis)
- 6th-7th Century AD – Bubonic Plague (estimated 25 million dead)
- 14th Century (1348–50 AD) – Black Death (estimated 75 to 200 million dead)

According to the above-mentioned timetable, the next event is due sometime in the 21st Century. Knowing that the Bubonic Plague took the lives of 25 million people and the Black Death took more than 75 million lives, we can only guess the number of victims during the next polar shift because the population of mice is in a direct ratio with the population of humans. The more cities we build, the more food and fewer predators they get.

Laboratory research proves that exposure to a strong magnetic field leads to sudden aggression in mice. If the hyperactivity of mice coincides with the period of these events (the siege of Jerusalem by the army of Sennacherib in 7th Century BC, the mass death of millions of people from the Bubonic Plague and the Black Death in the 7th and 14th Centuries AD), then we have to look for its source.

The next question that comes to mind is: if there was a sudden change in Earth's rotation, then why has nobody documented this phenomenon in the 7th and 14th Centuries? If the appearance of deadly diseases was coinciding with the polar shifts, then there would be no one to record this occasion considering that entire populations of cities were swept away. Death was taking the rich and poor, and the educated and uneducated classes of the population.

When illness hits us, our interest to the world surrounding us diminishes rapidly. Horrified by being infected with a deadly disease, it would be unimaginable to think about anything else when the first fear is where to get food and water that has not been infected by the plague for oneself and members of one's family. Besides, we do not know how the absence or overdose of the magnetic field can affect our consciousness and memory. There is a citation from the same *Worlds in Collision* book which I found very interesting:

In Egypt an inscription of the <u>Eighth Century</u> that refers to the moon disturbed in its movement, mentions incessant fighting in the land: "While years passed in hostility, each one seizing upon his neighbor, not remembering his son to protect." It was no different <u>700 years earlier</u>, in the days of catastrophes caused by Venus. At that time an Egyptian sage complained: "I show thee the land upside down; the sun is veiled and shines not in the sight of men. I show thee the son as enemy, the brother as foe, a man slaying his father." (Pages 165-166)

The same consistent pattern we can see here is the <u>Eighth</u> and <u>First</u> Centuries, <u>upside down land</u> and <u>mass insanity</u>.

If you recall, by placing an audio magnetic tape near a real magnet, we can erase all the content of the tape. During Earth's reversal and the change in direction of the magnetic field, all living things can be exposed to the magnetic field of the sun and lose their memory accordingly. We should not forget that, after all, we are memory chips, and only after the regaining of Earth's magnetic field do all the bio-organisms regain their consciousness but now with a big gap in memory. Maybe that is the exact cause of such a lack of information about events periodically taking place on Earth every six to seven centuries?

One of the most obvious examples of such phenomena is the story of the biblical Tower of Babel where all of a sudden the builders began to speak in different languages. I will try to retell the same story using different words: "All of a sudden, the builders of the tower lost their common language." In other words, the common language of the builders was suddenly erased. Such a drastic and mysterious disappearance of a formerly common language can be explained by the mass loss of memory. Of course it could have been achieved by some secret weapon of God's (biblical version), but it could also be the result of naturally occurring shifts of poles and drastic changes in the magnetic field of the planet.

Another indirect proof of our dependence on the magnetic field is magneto-therapy. Despite the absence of official medical recognition of its efficiency, people, for some strange reason, are using magnets from ancient times. Even though compared to Earth's magnetic field a small magnet is too weak for us to detect its influence, many sensitive people feel their power, and as the saying goes, "There is no smoke without fire."

Here is one more curious detail about Earth's recurring reversals. As we know, draining waters in the Northern and Southern Hemispheres flow in opposite directions. Of course it is impossible to prove or disprove water flow direction in the period before the recent reversal, but what can help us here are sea shells. As we know, sea shells grow in both directions. The direction of their growth cannot be independent from Earth's direction of rotation. All the living things on Earth live by the same cycles as Earth itself, and it would be logical to assume that the sea shells of the Northern and Southern Hemispheres have to have oppositely directed spirals.

"Imagination is more important than knowledge."

Albert Einstein

In this chapter I would like to take you into an imaginary time travel expedition into the history of our beloved Earth. We are on board of some complex time-travelling device and ready to head into the past. So here we go.

Through the darkness of endless space we begin to approach a small planetary system, and after a torturously slow travelling time, we see our beautiful and lovely planet, our homeland. As we are approaching, we can clearly see the stripes of white, blue-green and white again. But these are not the colours we were expecting to see; and besides, **where is our moon?** By getting closer and closer we begin to experience a constantly growing feeling that we are at the wrong destination even though all the sensors are indicating the right destination and time.

While revolving around the globe we can't identify any shapes of present time continents, just a green planet with multiple large arch-shaped seas (there were no oceans), fenced by continuous mountain ranges. This three-coloured planet consists of blue seas and green forests sandwiched between the two white polar caps, which seem surprisingly large to our eye. What catches our attention is that the planet we are visiting has no tilt whatsoever and is spinning at a much higher speed than in our time. It is revolving around its axis like a huge whirligig in space.

During our scheduled descent and preparation to land we can see a very green and friendly planet with its countless diversity of flora and fauna. But those are not the trees and animals we are used to seeing in our time. By gliding above the sea, we notice pods of dolphin-like creatures that are diving in and out of the water, accompanying some large ships. By flying right over them we are stunned to see the hairs and human torsos in them. Jumping back from our windows we look at each other with fear and doubt in our eyes. Did we just see humans on the decks? But the dashboard was indicating 15,000 years back.

During the flight above the mainland we notice that everything there has quite huge proportions. We see herds of countless mammoths feeding among huge sky-scraping trees with 200-300-metre-high peaks. The sight of enormous dinosaurs is a complete surprise to us because we were told that they were extinct millions of years ago, and all of a sudden here they are. Calmly suspecting a programming mistake, we try to absorb as many scenes as possible. We have a feeling of watching the Jurassic Park movie, but in more realistic dimensions.

After the most exciting cruise around the globe, we land on the outskirts of a huge metropolitan city with its buildings and parks. It is quite unexpected to see any man-made structures located so close to wild forests with their giant, multi-coloured inhabitants. At first, we try to draw a parallel between these mega structures and our own urban structures, but it is nothing alike whatsoever. First of all, the proportions of those structures are quite impressive, and it looks like somebody just shrunk us to the size of midgets. Along with this gigantic environment we noticed the unusually larger disk of the sun. It seems like someone has inflated our sun and soon it is going to explode.

After some short preparations and strange hesitation, at last we decide to take a small trip into this densely-populated megalopolis and take a walk among the crowded streets of that strange looking city. What calms us down is our invisibility because we are travelling in a partly transparent but parallel dimension. Besides, travelling through thousands of years and missing the opportunity of seeing our ancestors would be very regretful after all. After a short walk we are on the streets of a very busy and quite loud settlement. What strikes us right from the first moment is the diversity of human-like creatures. Along with people of enormous (for us) proportions, there are mythological centaurs and other, completely strange creatures.

The average stature of inhabitants is 5 to 6 metres high, and what is most intriguing is that everyone in this huge metropolitan settlement is busy with his own tasks without paying attention to the creatures that are completely different from them. We have a strange feeling that even if we were visible nobody would even pay attention to us. We have a hopeless eagerness to have a word with them about our existence, but it is an impossible task to accomplish. We are just viewers from a different epoch.

As we move towards the centre of the settlement, we see the peaks of the central castle. By the signs that show some importance, we realize that it is the palace of the Gods. Creatures on the streets of this busy megalopolis can't see us because we are just visual visitors from the future, but deep inside us we feel the anxiety of being seen by Gods. Some strange feeling of guilt is driving us away from them, like students who are trying to avoid the principal because of a missed class.

After a quick sunset (the planet had much shorter days and nights) we decide to ascend to near Earth's orbit to observe an upcoming great event — the polar shift.

Finally, after endless and torturous hours in space, we witness Earth's gradual wobbling. Like a giant train after its long, long journey it is preparing to stop at its last station. Because of the enormous waves, which were swinging the globe from side to side, it looked like the Earth was hesitating to choose locations of its future poles until after all, the globe tilted on its side, and its former poles appear at its equator. It reminds us of an Easter egg with painted white tips spinning in space.

Being anxious about the fate of the inhabitants of that megalopolis, we leave this epoch of our planet's history to return to share our excitement with our colleagues. Our intention is to return at some later time. We decide to set our "magic" time travelling device a few hundred years later, and after weeks of preparations we are ready to dive into the depths of time again. Our spacecraft-looking apparatus slowly takes off from the present time and starts its smooth voyage into the past.

And again, after a seemingly endless travelling period, the cloud-like matter around our apparatus begins to dissolve, and we begin to see some familiar shapes of the gradually approaching planet. By getting closer and closer we notice some strange changes there. The Earth is not equally coloured like before. The polar white colours prevail over blue-greens and the polar ice is unusually thick. The Earth is experiencing an Ice Age.

The seas we saw before have changed drastically. Instead of equally spread multiple seas, the planet is covered by one enormous ocean with numerous tiny water basins. The main water body of the globe is in the bucket between the mountain ranges. How the waters of the world's oceans could gather at one location remains a mystery to us. By continuing our voyage through the coming years, we realize that this enormous ocean is constantly growing. Like a giant vampire, it is sucking all the waters from others, and after centuries of constant growth it is spreading farther and farther. Such an "unevenly inflated balloon" is not looking stable at all. After our numerous revolutions around the globe, we realize that most channels escaping from the ocean are blocked by constantly growing polar caps and now, without any doubt, this monstrous "vampire" will spread even farther.

After our painful observations of this "time bomb" we decide to descend to Earth and see how people and animals are dealing with such unfortunate changes in climate. We land on the top of the mountain range serving as a dam that prevents the constantly rising ocean from spilling off. People living on the shores of the ocean are constantly building water dams to prevent their settlements from flooding. But after the continuous rising of waters they have no other choice but to move their homes to higher and higher grounds. The sight of people living in constant fear is quite unbearable, and we decide to see if anyone lives on the other side of the dam. And here, not surprisingly, people located on lower altitudes are also living in constant migration, but contrary to the people living on the shores of the Alpine mega-ocean, the inhabitants of the lowlands are constantly following retreating waters.

The very next day we decide to visit sites remote from the ocean and not surprisingly, we notice that places located away from that rapidly growing "monster" are getting dryer and dryer. Continuous droughts are eventually wiping out settlement after settlement. Hunger is gradually exterminating all the living things.

The first question that comes to mind is: "Why doesn't anyone try to ease the water pressure of that 'Vampire Ocean'"? Conduction of local explosions in these natural partitions could stop the growth of the ocean and prevent the approaching catastrophe. Along with saving the planet, it would save millions and millions of lives. Delaying it year after year was making the situation irreversible. With constantly rising sea levels, that monstrous ocean would be harder and harder to control, but what we notice here is that the majority of Earth's population has no idea what is going on, what their beloved planet is going through, and what catastrophic event is awaiting them in the future.

During our time-voyage around the globe into that epoch, we experience a strange feeling of Earth's foregone destiny. Why were the Gods who have travelled to Earth through multiple galaxies, who created humans and animals in diverse races, sizes and abilities, not doing anything to save the Earth? Didn't they have all the necessary power and capabilities? Even silent observation of planets neighbouring Earth seemed like all the members of our solar family were against our beautiful and fertile planet. Meanwhile, every day was bringing the Earth closer and closer to an irreversible catastrophe.

Due to our unwillingness to see all the pain and horror of Earth's inhabitants, we decide to leave that period and traverse to the time when all those torments would be over. When we visit the Earth right before its upcoming reversal, the situation has changed from bad to worse. Drought and hunger sweep away entire populations of equatorial regions. People who live close to seas are the most fortunate. With shrinking seas, all living creatures are migrating behind retreating waters, thereby providing a necessary food supply to the remaining population.

The "Vampire Ocean," on the other hand, swells to such an enormous size (it covers half of the Earth's surface), that from a far distance our planet looks like a pregnant woman in her last months of pregnancy. Considering Earth's unstable condition, any abrupt move would be crucial for it. The Earth is experiencing its last prenatal period. Time passes by and the inescapable moment finally arrives. The earth is experiencing its orderly transformation of shifting its poles. But this time, not everything is flowing as smoothly as before. When the Earth begins its complex "dance of reversal," where one of the poles wobbles around the second one, its egg-shaped globe begins to transform into an even more elongated object. The rapid tilt of Earth's axis makes the entire floating polar cap of the "Vampire Ocean" glide away from its original location, creating an enormous wave in front of it. This massive "bulldozer" pushes the waters of the Vampire Ocean with unprecedented speed and power until it just tears off from the rest of the planet along with a multi-kilometre layer of land. The sound wave of the explosion is so powerful that it reaches all the planets of our solar system. The energy released from the separation of that enormous chunk is so high, that more than half of Earth's surface is ejected into the space. The scene is so unexpectedly frightening for us to see that we cannot understand whether it is real or not, but right after that, the first picture which comes to mind is the picture of the splitting of cells from the school books. But could it be possible?

It resembles a horrible execution of scalp removal, where in a matter of minutes more than half of Earth's surface is torn away. Despite the separation of such a large layer from Earth, a spontaneous shift of land squishes the remains of the former "Vampire Ocean" so much that its waters instantly break through the barricades in two places and begin to rush down to inundate the rest of the planet.

Gradually, we begin to realize that it really did happen to our beloved planet. What is most awful to realize is that half of the Earth's crust, along with mountains, oceans, plants, cities, animals and people, all of them are ejected from Earth and melt down to one enormous sphere of fire which is spinning through space like a huge fireball. Spinning with unimaginable speed, it is moving farther and farther away from the Earth. Like a newly hatched baby-turtle, it is rushing away from its nest.

The separation of such a large portion of Earth's surface is accompanied by myriads of debris lifted by the "newborn baby." What goes up must eventually come down, and after reaching their maximum altitudes all the multi-size fiery balls begin to bombard our already weakened planet. Large spheres, fallen to the ground are creating enormous mountains of fire and others are splashing into the ocean, creating huge waves and massive clouds of smoke and steam. Torrential rain is pouring on the still boiling Earth only to evaporate again. This thick layer of grey fog is not letting us see the entire picture of that tragic event. The planet is chaotically tumbling from side to side like a giant wounded animal trying to stand on its feet.

Beneath the black-grey cloud that envelopes the Earth we see myriads of chaotically bursting red explosions. Meanwhile, one of the red, blurry spots on the surface of the earth begins to grow rapidly until it suddenly appears from the cloud as another massive, fiery sphere of molten lava. It seems that it appears right from the centre of the Earth and is heading straight towards us. Even though it has much smaller dimensions than its sibling, we can feel its fiery breath and multiple loud explosions on its surface. This time, unlike its elder brother, the newborn offspring decides to stay with its mother and begins to follow her in space.

As we watch this process from space we can't get rid of the thoughts that along with the agony of Earth there is a complete extermination of every living thing on it. It will be impossible to point out any relatively safe place on Earth to survive. Even being on the highest mountain peak of the remaining quarter of the land will not guarantee anybody's safety. Creatures who survive this unprecedented earthquake, intense heat and drowning, have to avoid the bombardment of the fiery balls and acid rain. And even after miraculously passing through all that hell, they have to sustain their lives with non-contaminated food and water. How could anything survive in such a harsh and sunless environment would be a complete miracle. Just the fact of our visit from the future was proof that our ancestors somehow did survive this cataclysmic event, and we can't find any other explanation for that survival except for it being a MIRACLE.

After a very exhausting and meaningful time-travel, we decide to leave our planet's past and return to the present time. Unlike our first trip, when we joyfully wanted to share everything we saw with our colleagues, this time we just want to keep silent and never return to those destructive times again. Our return trip was full of thoughts and was accompanied by absolute silence, where every single one of us was trying to analyze the scenes we saw.

In this short, historical narrative, I tried to introduce my readers to real events which took place on our planet Earth, not millions but only thousands of years ago. In the following chapters I will try to prove my vision, scientifically.

Lakes

Our planet is scattered with thousands and thousands of lakes. All of them are different in size and depth but the most important features of the lakes are their location, altitude and salinity. Therefore, all lakes have to be classified by their origins;

1. Lakes being trapped in lowlands of rising equatorial bulges
2. Lakes being formed by the melting of polar ice
3. Lakes created by the Great Deluge

Let us consider them one by one:

1. With hundreds of lakes scattered on their tops, a typical example of lakes being formed by rising equatorial bulges is the Himalayas. What is fascinating about them is that most of them are saline and at the time when the Himalayas were elevated to their current altitude, absolutely all the lakes were saline. They were just scooped up from saline oceans by piling equatorial bulges. We can classify those lakes as follows:

    a)   Lakes with outflow channels have been drained away right after their formation.
    b)   Lakes without external water infusion were gradually drying out, creating deserts of salt.
    c)   Endorheic (with no outflow) lakes feed up from spring waters and always stay saline.
    d)   Lakes which have both feeding and draining rivers eventually lose their salinity and become fresh water lakes.

One of the best examples of basins raised by equatorial bulge is the bowl-shaped Kashmir Basin located at 1,600m altitude. The presence of microscopic seashells scooped from the bottom of the sea, along with all the living creatures in it, proves the existence of the lake in the past. According to Hindu legends, a long time ago the lake was inhabited by a large sea monster. Constantly attacking and devouring people, it was spreading fear and horror to habitants who lived near the shore. Only because of natural drainage of the lake, was a local hero able to kill the monster. An animal capable of attacking and eating people should have quite an impressive size and — evolutionarily — the existence of such giant prehistoric monsters in a lake just does not make any sense. To grow to such monstrous sizes there should be plenty of food, and small enclosed water masses cannot provide such an abundance of food, which means that the animal was just trapped in a rapidly ascended bucket.

According to ancient myths and modern testimonies this is not the only mention of prehistoric creatures living in lakes. Monsters have been seen in lakes Loch Ness, Van, Baikal and so on. Here we should not confuse high altitude lakes with lakes located on lower altitudes. Lakes Bayial and Loch Ness have completely different origins. Those are just cracks in Earth's crust. Since transformation of salt-water lakes into fresh water lakes takes place gradually, all the salt-water creatures also gradually adapt to fresh water environments (sea horses of Lake Titicaca, for example). There is a very curious incident described by a missionary traveller, M. Huc, during his voyage to Tibet in 1846:

> "At the moment of crossing the Mouroui-Oussou, a singular spectacle presented itself. While yet in our encampment, we had observed at a distance some black shapeless objects ranged in file across the great river. No change either in form or distinctness was apparent as we advanced, nor was it till they were quite close that we recognized in them a troop of the wild oxen. There were more than fifty of them encrusted in the ice. No doubt they had tried to swim across at the moment of congelation [freezing], and had been unable to disengage themselves. Their beautiful heads, surmounted by huge horns, were still above the surface; but their bodies were held fast in the ice, which was so transparent that the position of the imprudent beasts was easily distinguishable; they looked as if still swimming, but the eagles and ravens had pecked out their eyes."

> M. Huc, Recollections of Journey through Tartary, Thibet [Tibet] and China during the Years 1844, 1845 and 1846. Vol.2, pages 130-131 (or page 235 in 2005 edition)

As we can see here, wild oxen were frozen alive because of sudden changes of latitude and altitude of the landscape they were situated on. Located on moderate latitudes, wild oxen had witnessed the globe's recurrent polar shift. The fact that these frozen animals were found on the highest mountains on Earth indicates the nearness of a giant polar cap in the past. Animals located in the equatorial region were just trapped in frozen water and then suddenly displaced to the pole. The following recurrent polar shift had displaced their frozen carcasses to warmer latitudes again, and due to the appearance of an equatorial bulge, the pond was elevated to higher altitudes to remain frozen forever.

2. A typical example of lakes being formed by the melting of polar ice is all the freshwater lakes of North America. There are myriads of tiny lakes and ponds scattered near the Hudson Bay Basin and on the Canadian Arctic Archipelago. Lakes located outside of that great circle are saline, and the salinity of the soil led to the formation of prairie regions. The presence of

fresh water in the soil has a favourable effect on the growing forests and the presence of salt in the soil leads to poor vegetation respectively. That's exactly what we have here: thick forests on the north and prairies on the south.

3. The lakes created by the Great Deluge. If, according to the Old Testament, there was rain for forty days and forty nights, then such an event could fill the highland basins up, transforming them into water reservoirs. Postdiluvian lakes with inflowing rivers have been preserved up until our days, while the lakes without water inflow have eventually dried out.

Ice Age

During each epoch, the accumulation of polar ice was gradually changing the globe's centre of gravity. After an x-amount of time when the volume of polar ice was reaching its critical point, the earth was careening on its side. If we look at our poles today, there are no obvious round shaped borders around them, and also there is not as much snow as during an Ice Age. Of course, in order to obtain a perfectly circular shape, a pole has to be located either completely on land or in the middle of the ocean. But there is also another reason. Thanks to the changing of seasons, our geographic poles never get so much snow accumulation to cause enormous dents on Earth's surface, which means that back in time there was something completely different about our planet. Here we can build a logical chain of thought:

1. All arc-shaped features on the globe were indented by enormous masses of ice.
2. The only condition for ice to accumulate on such a massive scale is the occurrence of an Ice Age.
3. For a massive Ice Age to take place, the globe should not have recurrence of seasons, which means it should not have any substantial rotational tilt.

4. The absence of a rotational tilt means an absence of any satellite.

If we remember, the Garden of Eden had no alternation of seasons. Even though there was "eternal spring" on 45° latitudes and a complete dry out of equatorial regions, the planet was experiencing a major Ice Age on its poles.

The existence of so many arc-shaped mountain ranges on Earth indicates a frequent occurrence of Ice Ages in the past and that fact proves that at some point in time our Earth did not have any satellites around it. Due to the absence of a rotational tilt, poles located on dry land have been accumulating snow very rapidly, causing an enormous indentation on the surface of the globe. After reaching its apogee, every recurrent reversal was shifting the poles to the new locations. In other words, poles constantly "walk" on the surface of Earth.

Presently, thanks to our satellite, the Moon, and the existence of ocean currents we have moderate climate on the planet. Our moon plays a role of "equally heating the globe grill" and ocean currents, on the other hand, it can be compared with constantly stirred soup. The largest ocean current on Earth today is the powerful Antarctic Circumpolar Current, which flows clockwise around the Antarctic continent with an average speed of 4 kilometres per hour transporting an enormous volume of water. It would be hard to imagine the consequences of its complete stop.

According to scientific data, the last Ice Age ended with the Great Flood approximately 10,000-14,000 years ago. But what is an Ice Age? An Ice Age means extended glaciations of geographical poles, and here are common conditions for its occurrence:

1. A supercontinent is located on a pole
2. Landlocked polar seas
3. A supercontinent covering most of the equator

After every recurrent reversal, each of these conditions can be applied. It is like throwing a dice where you never know what number is going to come up next. But because of the planetary tilt and recurrence of seasons, all these examples can lead to the occurrence of Mini Ice Ages. For a really severe Ice Age to take place, the planet should not have any tilt towards the sun.

Besides those three catastrophic scenarios, there is another, the fourth and a most lethal location of the poles, where all three conditions are met. A supercontinent happens to be located on the pole, the opposite pole is in a landlocked ocean and the other supercontinent covers most of the equator. If in addition to that the ocean is surrounded by high and continuous mountain ranges, then there is a big, big problem. If there are no escape channels, the waters of the globe will be gradually pumped into one landlocked reservoir, causing uneven distribution of global weight. The only considerably painless relief from this stressful situation is the planet's peaceful careening on its side. Of course, emerging mega-tsunamis would sweep entire continents from the surface of the globe, but that would be the less painful scenario.

Strange Geography

According to modern science, once upon a time "200-300 million years ago," there was one supercontinent on Earth. They even found a name for it: Pangaea. It sounds like the earth was resting on three whales, and then all of a sudden the whales decided to split up. But why would that one supercontinent gradually disintegrate into pieces? The only explanation for that gradual disintegration would be the gradual expansion of the planet itself. But is that what happened?

In the beginning of the 20th Century, the theory of a supercontinent and drifting continents was introduced by Alfred Wegener, a German researcher and geophysicist. Later on in the middle of the 20th Century, thanks to the development of paleomagnetism (rock magnetism), his theory was widely accepted by scientific societies around the world. But is it possible for magnetized rocks to sustain their magnetism for 300 million years?

According to science, in an ideal environment, a permanent magnet can sustain its qualities for a very long time. But how long is "a long time"? For us humans who started to use the magnets just 300 years ago, predictions of longevity of magnet properties can be very approximate. Additionally, why do we think that we live in an absolutely constant environment? We can only guess the number of polar shifts during those so-called 300 million years.

What is amazing here is that according to science, the process of separation of continents started some 200-300 million years ago. These unimaginably large numbers came from the discovery of radiometric dating methods, which are still the main dating techniques at present. I'm not a scientist, but I can certainly assert that the **effectiveness and precision of radiometric dating techniques stops at the threshold of the most recent major cataclysmic event.**

One thing Alfred Wegener was absolutely right about is that at some point in Earth's history, the American continent was joined to Eurasia and the African continents. The similarity of their coastlines is quite evident. But are they drifting away from each other? Scientific observations indicate that they are, and it happens at an approximate rate of 1-10 centimetres per year. But continents cannot just split into pieces without any substantial cause. For such a process to take place there should be the involvement of an enormous force.

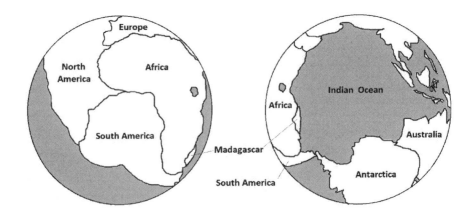

(Pic.2)

Having obvious pieces of a puzzle we can easily restore the look of our globe before the continental split. Virtually joining the American continent to the African continent *(Pic.2)*, we can see that two former poles (as proposed by Hugh Brown-Sudan Basin and Amazon) were in fact parts of the same enormous ice dent.

Another obvious and recent footprint of the former polar ice is the Hudson Bay Basin in Canada, where only a severe Ice Age with its enormous weight could leave such a large footprint on Earth's surface. Let us look at the east coast of the American continent. South America has a very distinct coastline. In contrast to that, the North American east coast looks very shabby. The exact same picture is on the west coast of Europe. The African continent has a very distinct coastline and Europe in fact is one enormous archipelago.

Let us reconstruct the locations of continents and poles on Earth before their separation. First, the North and South American continents along with Greenland were joined with Eurasia and Africa.

In Europe, the Mediterranean Sea did not even exist yet. The British Islands and Scandinavia were part of one united continent, including Asia, Africa and the entire American continent.

Instead of the Indian Ocean, there was a continent along with its seas and rivers. The so-called continent of Lemuria was covering the entire Indian Ocean, but without the so-called drifting from the Madagascar Indian Plateau."The Indian plateau did not even exist back then. I would mark the island of Madagascar as point zero and instead of the common opinion of it drifting away from the African continent it is the other way around. The African continent separated from Madagascar and the rest of Asia, creating the Red Sea and the Persian Gulf.

Vast areas between today's Kamchatka, Japan, Indonesia and Australia on the west, Antarctica on the south, Andes and Cordilleras on the east were also covered by a continent *(Pic.3)*, but because of being surrounded by high mountain ranges they were gradually transformed into an enormous ocean the size of the current Pacific and Atlantic Oceans combined because at some point in time the American continent was joined to Europe and Africa.

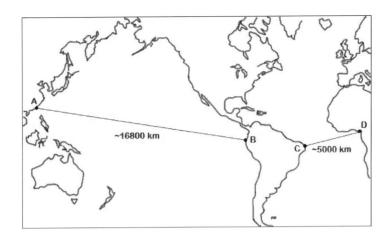

(Pic.3)

Let us compare today's Pacific and Atlantic Oceans. The first feature to catch our eye is the size difference *(Pic.3)*. The second and more important feature is their coasts. Compared to the Atlantic with its gradually tilting coasts, the Pacific Ocean has very sharp and abrupt coasts. If we draw a line along those steep landscapes of the ocean bottom, it will coincide with the "Ring of Fire" — the concentration of the majority of Earth's volcanoes.

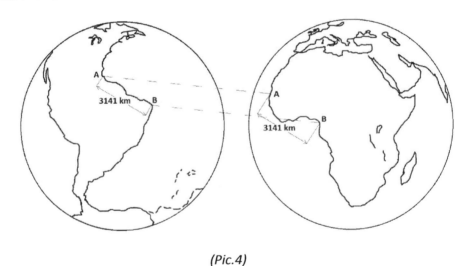

(Pic.4)

When we try to join the American continent with the African continent, we detect some curious detail. The reconstruction of continents by joining them together reveals some missing chunks of Earth's strata *(Pic.4)*.

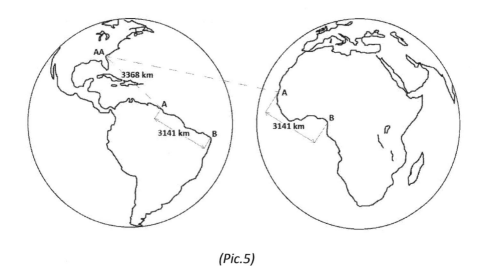

(Pic.5)

If we mark the joined points on the eastern shore of South America and the West African shore, we can clearly see that point (A) in South America does not coincide with the same point (A) on the North American east. When we try to join the American continent to Africa there is approximately a 3,368 kilometre discrepancy between North and South America *(Pic.5)*.

(Pic.6)

For the American continent to wrap around the present European and African continents, the North and South American continents must be pushed towards each other, eliminating the present Caribbean Sea and

Gulf of Mexico *(Pic.6)*. And that argument excludes the common opinion that the Gulf of Mexico is the result of asteroid impact and proves its appearance through the process of stretching of the two American continents.

"Vampire Ocean"

In this chapter of the book, I have to start by mentioning the moonless Earth. The planet which had been revolving in space without any (or very insignificant) tilt, and had just two completely opposite climates: minimally exposed to sunlight poles and constantly "frying" equatorial regions. The only regions with conditions moderate enough for life would be the regions between the equator and the polar ice. Basically, the planet would be divided into two corridors of life centred at 45° latitudes of the Northern and Southern Hemispheres. It is hard to deny any possibility of life on the equator, but one thing is obvious: the diversity of living creatures in equatorial deserts would be significantly less than in areas farther away from it. During an Ice Age, the situation in equatorial regions would be even worse because of rapid evaporation with no precipitation due to its retention in cooler areas of the globe.

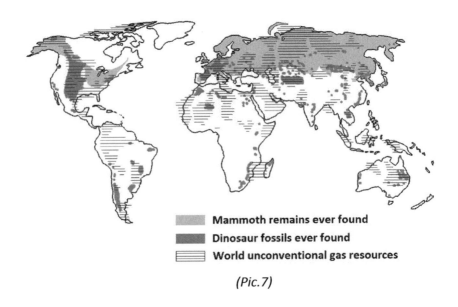

■ Mammoth remains ever found
■ Dinosaur fossils ever found
☰ World unconventional gas resources

(Pic.7)

From ancient times, people have constantly unearthed remains of mammoths and mastodons, fossilized bones of dinosaurs and other prehistoric animals. As we know, the entire flora and fauna of our planet consists of organic matter, and every dead corpse must be decomposed or be eaten by scavengers. Just the fact of the existence of fossils and remains of prehistoric animals points at a drastic catastrophe, where none of the basic laws was in effect. Instead of being decomposed, the corpses were frozen or fossilized forever, without even being touched by scavengers whose absence means a complete extermination of every living thing on earth. So much organic mass could not just disappear. Natural gas, oil and all the flammable substances are products of the process of decomposition, and according to the volume of current shale resources, there was decomposition of an unimaginably large volume of bio organisms.

Nowadays, thanks to documented discoveries of mammoth remains, dinosaur fossils and global shale resources *(Pic.7),* we can notice an important feature. The map above does not depict the precise location of all the findings, but what is important here is that data of all three maps indicate that the Hudson Bay Basin, the Canadian Arctic Archipelago and Greenland are free from any animal and vegetation remains, a fact which certainly leads to the absence of shale resources accordingly. This map provides clues of exceptional importance.

During the global cataclysm, the Hudson Bay Basin, Canadian Arctic Archipelago and Greenland were covered by an enormous ice shell, which prevented the deposition of any organic matter. At some point in time, right before some planetary-scale cataclysm, the Hudson Bay Basin was located at a geographical pole.

If we suppose that reversals never occur, then the planet would never exit any Ice Age stage and the constantly accumulating polar ice would cause the equatorial parts of the planet to dry out completely. Knowing the mechanism of recurrent polar shifts, we can create an exact chronology of events. So, at some point in time, after recurrent reversal, our planet ended up in the position where the present Hudson Bay Basin was moved to the pole located on dry land, where polar regions began their rapid accumulation of ice.

On the other side of the planet, the opposite pole was located either on water or on land. Nevertheless, even being situated on dry land, it had one major difference: the hemisphere it was located on was surrounded by high mountain ranges. Waters precipitating away from the pole were flowing to the lowlands of the landscape, and because the pole was in an enclosed area, all its rivers were feeding the same enormous basin.

Due to the absence of a tilt, the equatorial part of the planet was receiving intense heat, and polar areas, on the other hand, were experiencing extremely low temperatures. Rapidly evaporating waters of equatorial regions were precipitating back with one big difference: they were not flowing back into the global water system and were staying in the same basin, creating an ocean monstrous in size. Icebergs constantly splitting from polar caps were feeding this "chronically thirsty" ocean. The pole was essentially acting as a giant pump extracting every drop of water from the air. The air it was using as fuel was not collected solely from the surface of the "Vampire Ocean." It was gathered from the entire globe.

At some point, the constant shrinking of oceans led to the abruption of any bonds between the oceans. A complete stop of the already weak sea currents, the vital arteries of our planet, accelerated water drainage even further. All the equatorial deserts were now growing towards the poles. During hundreds of single-seasoned years of water pumping, constantly growing ice caps were gradually draining all the equatorial seas until there was nothing else to collect from beyond the borders of the now "Great Vampire Ocean." At the time of its peak, there was almost nothing to collect from the rest of the globe. Equatorial zones of the planet behind the ocean were transformed into deserts.

If we consider that Earth back then was much closer to the sun than it is now, then equatorial regions of the globe would not be fit for living, and people and animals would be forced to move away from equator in search of more moderate climates. One can only imagine the scale of an ever-proliferating global drought, which had wiped out entire populations of flora and fauna on Earth. People and animals that did not migrate away from the equator were destined to die from thirst and starvation. Hopes for an eventual return of torrential rains were in vain because the only possibility for that to happen would be either breaking off the mighty walls of the "Vampire Ocean" or the recurrent reversal of the globe. Both scenarios would end up with a worldwide deluge regardless.

Despite devastating drought and hunger on one part of the planet, it was a completely different picture on the shores of the ever-growing "Vampire Ocean," where the main problem was the constantly rising levels of seawater. The ocean was gradually swallowing every settlement and port on its shores.

In his book *Built Before the Flood*, Hans Bellamy scrupulously describes the "Inter-Andean Sea" ("Vampire Ocean") right before cataclysmic events. Back in 1943, during World War II, Hans Bellamy came so close to solving the Tiahuanaco problem that it would be very and very rational to scrutinize his works in today's schools. According to his research and latest archaeological discoveries, the underwater area of Lake Titicaca is covered by some prehistoric structures. That important fact answers the question of the creation of Lake Titicaca itself. The existence of ancient structures on its bottom leads to two possible scenarios of its formation:

1. Before the appearance of the "Vampire Ocean," the lake was much smaller than it is now.
2. Before the appearance of the "Vampire Ocean," the lake did not even exist.

After the gradual approach of the "Vampire Ocean," Lake Titicaca eventually became part of one enormous water basin. People were building new settlements and docks in areas not yet swallowed by constantly upcoming ocean waters. Every day they would witness the swallowing of their former houses by an ever-growing mass of water.

In his book *The 12th Planet*, Zecharia Sitchin describes gods who came to planet Earth with one intention: to mine for gold. That precious metal was desperately needed to save their mother planet, Nibiru, which was losing its atmosphere. For gods, who were capable of travelling from planet to planet, digging a water canal to balance our planet's oceans would not be so hard. If, according to the author, all attempts to mine gold from water did not get any results, then maybe stripping of vast land areas of the planet would be even more preferable for them. Maybe it would be logical for them not to undertake any measures to prevent the enlargement of the "Vampire Ocean." In order to answer that question, we have to try to reconstruct the chronology of events *(Pic.8)*.

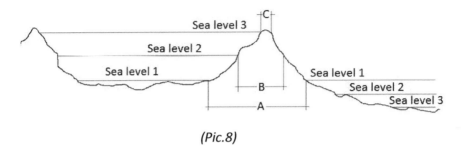

(Pic.8)

When everything just started and water levels were at Sea level 1, distance A could be equal to 10,000 kilometres, which is one fourth of Earth's diameter. To dig such a vast distance, even for gods would be quite a difficult task to complete. As we can see from the diagram, the rising level of the "Vampire Ocean" was shortening the distance from A to B and C, which seemingly would be less and easier to dig. But higher water levels mean higher pressure on the surrounding mountains, which makes it more dangerous to conduct any digging. It is like a stalemate in a chess game; whatever you decide to do is unacceptable. So we cannot blame our creators for things they could not control. (Here we can't completely eliminate the possibility of God's involvement in all these calamities. Artificial reinforcement of the weak spots in surrounding the "Vampire Ocean" mountain ranges would increase the probability of the future catastrophe).

Unfortunately, some cataclysmic event eventually demolished this enormous mountainous range and waves of the disintegrated "Vampire Ocean" swept entire populations of flora and fauna from the surface of Earth towards lower altitudes of our globe. During the following centuries, organic matter gathered in one area decomposed and eventually turned into fuel. All the gas and oil resources we are burning today are nothing but debris formed from animal and vegetation remains, piled up around the globe by waters of the "Vampire Ocean."

With great sorrow we have to admit here that this list also includes the remains of humans, the remains of people who used to live right before this terrible event, the remains of our ancestors. It sounds very disturbing, but the economy of our present civilization largely depends on mass graves of spontaneously exterminated people, as well as flora and fauna, who were flourishing on Earth way before our epoch. Just imagine thousands of years from now, future generations of midgets would scream and yell at their "stock exchange," selling decayed products of our remains. What a sacrilege!

Prehistoric Andes

Many of us know about the megalithic walls of South America, but if we are impressed by massive stones perfectly fitted together, then those are nothing more than some kind of concrete mixture. To shape each stone to perfectly fit the shape of the neighbouring stones would take an unreasonably long period of time, but that would be only in the case of using a hammer and chisel. Just the look of those puzzle-like blocks indicates that they were done by pouring some liquid mixture on top of the already formed blocks. When our explorers say that they cannot even stick the blade of a knife between those blocks, I could say that when we pour fresh concrete on the top of already hardened layer and after it hardens, we also cannot stick a blade between them. Then why does this fact surprise them so much?

A   B

(Pic.9)

Our ancestors used metal anchors to connect them together, and we presently also use metal armature in concrete. And the fact that the metal they used back then does not rust is thanks to the completely different environmental conditions they were melted in, and not the result of extraordinary technology of our predecessors. We should not forget that we are <u>not</u> talking about the planet we know today. It was a planet with a completely different climate, rotational period, distance from the sun, level of oxygen and magnetic field.

Most of the megaliths around the globe have one distinctive feature: the bulging out lumps located on the bottom of the stones *(Pic.9)*. What we can observe on the picture on the left *(A)* is that the corner blocks have a pair of lumps on each side which means that during their construction, in order to prevent leakage of the mixture from the formwork, exterior walls had to be secured properly to withstand the pressure of the liquid mixture. Connecting two opposite sides of the form with anchors would be the only solution to prevent the mixture from leakage. After the gradual hardening of the mixture and removal of encasement, pulling out all the connecting anchors would lead to a partial leakage of the viscous mixture, and that is the reason for such distinctive breast-shaped convexes on the bottom of each megalith.

The illustration on the right *(B)* shows that the connecting bolts had quite an assortment of shapes: round and square. But what is important here is that all connecting bolts were hollow. Such a concave shape of the lump would only be possible to achieve by pulling out the hollow object. But let us forget for a moment about the methods and sizes of the blocks and try to understand why our ancestors would build anything like that in such high altitudes.

The landscape of the area around the present Lake Titicaca and its altitudes has a little story to tell *(Pic. 10)*. In his book *Built Before the Flood*, the great researcher Hans Schindler Bellamy (1901-1982) studied traces of a much larger sea that formerly existed in the basin of present Lake Titicaca. He named it the Inter-Andean Sea.

And indeed, what we can see here is an obvious shape of a prehistoric sea created by a rising equatorial bulge. By piling up all the rocks from the bottom of the ocean, the equatorial bulge scooped a part of the ocean itself, creating a high altitude saline lake. The flattened bottom of the basin could be attributed only to the existence of a body of water, where all the sand, dust and salt were evenly and smoothly deposited on its bottom. The former lake and now perfectly flat saline desert of Salar de Uyuni is the perfect proof of the existence of salty water in the past.

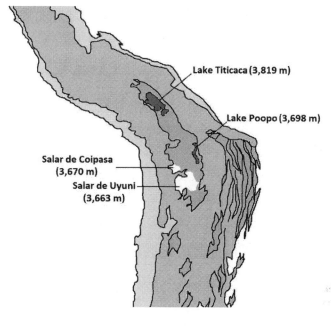

(Pic. 10)

As one of the best-preserved prehistoric sites on our planet, the megaliths of South America deserve very scrupulous attention and here, in attempts to unravel the past, we must examine them one by one and link them together into one puzzle *(Pic. 11)*.

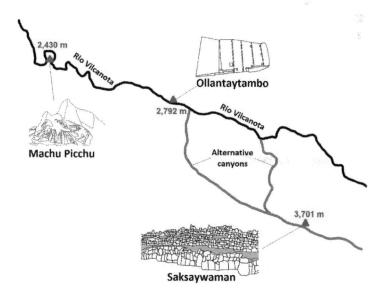

(Pic.11)

Machu Picchu, Cusco Region, Peru. Altitude – 2,430 metres. Located at the peak of the mountain, Machu Picchu is surrounded by the Vilcanota River, encircling it from three sides. If we consider that this same canyon was carrying waters from the Inter-Andean Sea (the "Vampire Ocean") into the present Amazon River, then it has all the necessary qualities to serve as a transit port connecting the high-altitude sea with the lowlands of the rest of the world. The comparably flat plateau on the peak of the mountain could have been achieved artificially.

By moving up the stream we get into the ancient megalithic site of **Ollantaytambo, Cusco Region, Peru. Altitude – 2,792 metres**, located in the canyon of the same Vilcanota River. The ancient site of Ollantaytambo can be considered as a reinforcement of the canyon walls.

The Megalithic complex of Saksaywaman, Cusco Region, Peru. Altitude- 3,596 metres. The building of such massive and monolithic walls could not serve as a fortress in a war against others. People who built this complex had more logical reasons to spend so much time and effort. The zigzag shape of the complex was needed to withstand the almighty waves of the Inter-Andean Sea (the "Vampire Ocean"). Besides its role as a sea wall, it could accommodate many wharves on it. The multilevel character of the complex indicates a gradual sinking of the lower levels and the necessity to build more and more levels of protection.

Kalasasaya, Bolivia (Altitude 3,863 metres) is one of the archaeological sites of Tiahuanaco (Tiwanaku), located just south of Lake Titicaca, Bolivia. Constructed mostly from monolith blocks, they were built as seaports where one of the wharves could accommodate hundreds of ships (**not just boats, ships**). Taking into consideration that the Inter-Andean Sea had a much larger size than present Lake Titicaca, the building of such large ports on its shores is quite plausible.

The Archaeological site of Tiahuanaco *(Pic. 12)* is located some 16 kilometres south of the southern end of present Lake Titicaca. Since its discovery in 1549 by Spanish conquistador, Pedro Cieza de Leon, many looters and amateur archaeologists visited the site.

For four centuries, nobody could explain who built these megalithic structures and why until the middle of the 20th Century when German archaeologist Edmund Kiss suggested that the symbols on the so-called "Gate of the Sun" could represent an ancient calendar and that calendar could depict the times of the world that flourished here way before our own. It was a green light for many archaeologists and astronomers to look at the calendar of Tiahuanaco from a different perspective. And after many expeditions to the Tiahuanaco site, in 1956 Hans Schindler Bellamy and Peter Allan published their great scientific work, *The Calendar of Tiahuanaco*. It is a great work conducted by great scientists who scrupulously investigated the mysterious symbols of the so-called "Gate of the Sun."

(Pic. 12)

After years of research they concluded that this archaeological site represents an ancient calendar. But what is noteworthy in it is that one Earth year back then consisted of 290 days, and the Earth itself was moonless.

By accepting the originality of the 290-day calendar of the "Gate of the Sun," we can conclude that our planet was much closer to the sun than it is now. Closer to the sun means warmer climates. If the present year consists of 365.25 days and one day equals 24 hours, something tells me that a 290-day calendar year should have fewer hours in it because the absence of a moon means the absence of a slowing force accordingly.

Warmer climates and the lavishness of water and oxygen were promoting the plants to grow higher and higher thereby thickening the layer of oxygen in the air. Shorter rotational periods on the other hand mean a stronger centrifugal force, a stronger magnetic field and a stronger immunity of the planet accordingly. All that had a beneficial effect on the size of all living things on Earth.

Here we can certainly assert that the attempts of our genetics to create real size giant prehistoric creatures would be a complete utopia. There would be a shortage of one of the most important ingredients: the Earth at the time of giants. Unfortunately, that golden era has gone forever, and Earth had to step into the next level of its existence.

The First Great Catastrophe

Knowing the location of one of the poles (the Hudson Bay Basin), now we can traverse to the second pole of the prehistoric Earth: the "Vampire Ocean's Ice Generator." What we have there today is just a vast ocean with an obvious scar of a piece torn away from Antarctica. Here we can make an assertion that the second pole was in the ocean or very close to it because for water to fill up such an immense reservoir, there would have to be a continuous supply of melting icebergs.

At the time when Earth approached its reversal point, the "Vampire Ocean" was at its peak with the highest sea levels and volume. It had collected every available vapour and now a time of relative stability had begun. It was a time when the growth of the "Vampire Ocean" began to slow down. Much like adding a counterweight on the scales to achieve their balance, the earth was passing its last stage of stability and every litre added to the "Vampire Ocean" could play a fatal role. All equatorial regions of the planet located on the opposite side of the ocean were transformed into lifeless deserts, and the enormous ocean had no more resources to levy. People living on the shores of the mightiest ocean on Earth could at last take a break from the continuous struggle with upcoming waters. But, unfortunately, everything eventually ends and that fatal litre achieved the required balance, and the Earth began to wobble.

The beginning of Earth's reversal could lead to major floods on the shores of the "Vampire Ocean" located in the direction of the present sunrise, but the main force was still to come. Knowing the fact that during a reversal, only one of the poles conducts the "belly dance" around the other pole, we can easily determine the pole that played the fatal role in our planet's havoc. The centrifugal force always pushes objects away from its axis of rotation, and the same rule applies to polar floating ice. That's right, floating ice. And here we can make a major assertion with the following logical chain of arguments:

1. The only force capable of splitting the continents apart is a pulling force suddenly tearing away enormous chunks of Earth's crust.

2. For such a tremendous pulling force to appear, the surface area of the spontaneously splitting layer of the planet must have more than half of the entire surface area of the globe.

3. The only thing on Earth capable of gaining such momentum is water.

4. The only water stirring force capable of creating such an enormous wave to obtain the momentum of ejection is the floating ice of the poles.

5. The only possibility for floating polar ice to move from its resting place and accelerate away is a sudden polar shift.

One of the most important conditions for floating polar ice to attain such tremendous momentum is the altitude of its location. The increased mechanical arm of throwing movement played a vital role in the tearing off of a huge part of Earth's crust. It is equivalent to throwing a ball, where we need to stretch our arms out and apply force. To throw the ball with folded arms will yield much lesser results. We have an absolutely identical move here.

The second, quite important condition for a floating continent to blast off is the existence of a substantial barrier on its way, be it a continent or even a large island. Like a vehicle crushing into a heavier barrier will experience a sudden lift off of its tail, bumped into (substantial by weight) land, the floating continent of ice also would acquire a sudden lift off of its tail. But there is one major difference between these seemingly identical moves:

- An instantaneously elevated tail of the impacted vehicle will fall back again.

- An instantaneously elevated tail of an impacted to ground "iceberg" would continue to rise due to the sudden conglomeration of water behind it. Such a massive lifting force would lead to the rolling up of a formerly flat surface of ice and land *(Pic. 13)*.

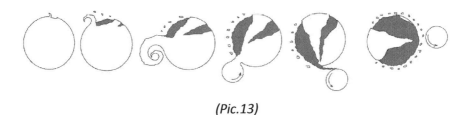

(Pic.13)

Floating polar ice in the "Vampire Ocean" had all the necessary conditions to break away: immense mass, high altitude, gradually accelerating wobbling of the globe, the enormous runway of the "Vampire Ocean," and in addition to all that a sudden barrier on its way.

Now, let us imagine our planet Earth having a tremendous womb, the "Vampire Ocean," with an enormous floating iceberg in the middle, which begins to travel away from its original location with increasing velocity. In a matter of minutes, having immense size, mass and thickness, the polar cap began to accelerate towards the eastern coasts of present-day Asia. During the rapid acceleration of floating ice towards the globe's equator when the linear velocity of the object reached its maximum speed, a sudden barrier of solid continent on its way may have forced the floating continent of ice to pile up on the point of impact and to acquire the necessary momentum to blast off from the surface of the earth. *(Pic. 14)*

Ejected into the space, the continent had created an enormous tidal wave that pulled no less than half of the planet's surface behind it. The emitted kinetic energy was so immense that along with the ejected continent of ice an entire multi-kilometre layer of stratum beneath the "Vampire Ocean" was also torn away from the surface of Earth.

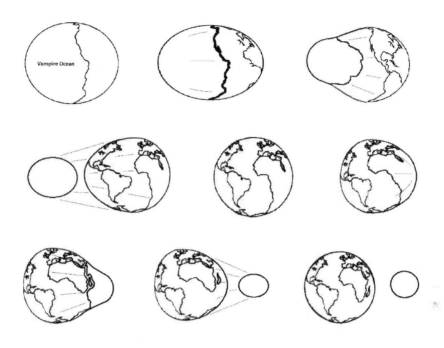

(Pic.14)

The tearing off of such a large chunk from the surface of the Earth drastically deformed the formerly spherical globe into an egg-shaped object. After the complete separation of the (now autonomous) satellite, the most outstretched part of the globe would rush down towards its centre in order to shape a sphere again. Considering such an enormous altitude of a portion of the globe elevated behind the ejecting satellite, its sudden retreat created an unprecedented force directed towards the interior of the globe. After its impacting back to Earth, it appeared that a tremendous wave had travelled around the entire surface of the globe to meet on its opposite side. It collided in one spot, and was forced to blast off of another smaller satellite into the space. And in order to have a complete picture of those catastrophic events, we have to remember that during this entire process the globe was chaotically spinning in space, creating enormous tsunamis, and myriads of other debris were spontaneously ejected into space.

Waves destroying everything on their paths were carrying gigantic debris of already lifeless remains of flora, fauna and people. Gradually subsiding tsunamis left tons of rapidly decaying organic matter scattered around the globe. The simultaneous decaying of such large masses of organic matter would require the consumption of an enormous volume of oxygen where the absence of a reimbursing it back mechanism (photosynthesis) led to the drastic reduction of oxygen in the atmosphere. The level of oxygen was also reduced by the process of rock formation because petrifaction of rock was accompanied by the trapping of air in it. Here we can certainly assert that after the ejection of satellites, Earth's atmosphere lost a significant amount of oxygen.

The Water

Considering that present oceans cover 71 percent of our planet's surface and knowing the process of satellite creation, the first question that comes to mind is, if a substantial portion of the "Vampire Ocean" was blasted off from the surface of Earth, then where did all the waters of present oceans come from? Here we could consider three plausible explanations.

1.   Considering the volume of satellites ejected into space, we can surely assert that the size of Earth before the ejection of satellites was much greater. But a greater size of the planet is not characterized by the volume of its hard matter only; it is also bigger by the volume of water, which means that volumes of world oceans of the pre-satellite ejection period and post-satellite ejection periods cannot be the same. The most comprehensive explanation of that process can be made by comparing our globe of the pre-satellite ejection period with many dairy products of our everyday life. That's right. Be it sour cream or yogurt, notched into any viscous mixture, pits will eventually be filled with water. Basins filled with water of the present-day Pacific and Atlantic Oceans are a result of such a process. The waters seeped towards the planet's lowlands from the entire globe. It does not mean that before the ejection the entire planet was covered by oceans; it means that before the ejection of satellites the entire crust of our globe was much softer, just like fresh sour cream. The ejection of substantial chunks of the globe led to the gradual seeping of world

waters to lower altitudes, leaving the highland regions of our planet with no moisture. The following centuries of waterless environment had eventually hardened the formerly soft tissues of the globe. It sounds strange, but the mountains we are admiring today are just scars of the healed wounds of the globe.

2.   Besides the above-mentioned filling of lowland basins by the gradual drainage from higher altitudes was the involvement of another water generator. As we know from chemistry: hydrogen + oxygen = water. The only condition for this reaction to take place is fire. Having little idea about the existence of free molecules of hydrogen in a pre-satellite ejection period we can assert that the percentage of oxygen molecules before the ejection of satellites was much greater than after the ejection, and the ignited planetary scale spark could create a colossal volume of water. The coexistence of oxygen and hydrogen together can be seen in the atmosphere of Mercury where despite the 427°C heat, 22 percent of hydrogen molecules do not combine with 42 percent of oxygen molecules, which may be due to such factors as a weaker magnetic field and atmospheric pressure accordingly.

3.   As we know, planets surrounding a star are not autonomous bodies, and streams of charged hydrogen ions (protons) are constantly bombarding the planets of our solar system. It is hard to say if by penetrating our atmosphere, these "solar messengers" react with free molecules of oxygen to form water or not, but they can definitely lower our water's PH.

The Scars of the Earth

After ejection of the satellites and going through all these calamities, Earth's climate had drastically changed and that process could not pass without a major impact on every living creature on it. Whoever survived the destruction and flood had to survive upcoming starvation and thirst, diseases and attacks of ferocious animals and degenerate people.

But besides the above-mentioned chaos, there was an irrevocable change in the magnetic field and rotational speed of our wounded planet. All the giant species of animals along with races of giants had lost their former power. In a matter of hours, their muscles were turned into a heavy and disobedient mass of meat. Creatures who miraculously survived this cataclysm were destined to starve to death or be eaten by smaller predators. Formerly almighty species of giants were turned into the most vulnerable creatures on Earth. The extinction of super giants during the First Great Catastrophe can be marked as the beginning of an era of much smaller creatures. Knowing the catastrophic events which took place on our planet thousands of years ago, we can reconstruct the exact chronology of events:

- The first stage we must start with is the creation of the "Vampire Ocean." The Earth's previous reversal put the planet into a position where one of the poles ended up in the location of the present-day Hudson Bay Basin, which was located somewhere near today's Ireland. Considering that the Hudson Bay Basin was located on dry land, it began to accumulate snow rapidly, pushing the planet into an Ice Age.

- The second pole was located entirely in the ocean where large icebergs were constantly breaking off from the main mass of the icy continent and, thanks to the centrifugal force, were carried towards the equator. Considering its landlocked situation, waters flowing into the same ocean were gradually expanding the borders of the ever-swelling "Vampire Ocean."

- The gradual and uneven distribution of weight around the globe eventually led to its polar shift. Due to a great imbalance of the weight between heavier polar regions and the equator, the globe careened on its side with great velocity.

- The spontaneous displacement of floating polar ice from its original location to the equator led to its ejection from the surface of Earth.

- Rapid retracting of the stretched-out part of the globe towards its centre led to the appearance of an enormous bulge on the opposite side of the globe, and its following blast off (the crater of the present-day Indian Ocean).

I don't have the intention to rename our strange calendar's numeration, but if we want to link ourselves to our planet and its prehistory, I think we must use the following terms: "Earth before the ejection of satellites" and "Earth after the ejection of satellites." Located on the northern outskirts of Cusco, Peru, is a torn-away chunk of a formerly grandiose structure: the megalith of "Upside-down Stairs of Sacsaywaman." This definitely belongs to the epoch before the satellite ejection because only the abrupt displacement of an entire continent could tear and carry such a huge monolith away from its original location.

Thanks to the fact that many Andean peaks have never been reached by the "Vampire Ocean" and the following global cataclysm did not annihilate this region, many types of plants avoided the sweeping away catastrophe. No wonder so many varieties of corn, beans, potatoes, cashews, tomatoes and other plants were added to our daily rations after the rediscovery of the American continent. It is hard to imagine our present cuisines without the miraculous survival of all these plants.

The separation of the main satellite led to the tearing off of the pole-to-pole slice of the globe (the entire American continent). During its spontaneous 5,000 kilometre displacement, the strongest tension of the pulling force was concentrated in its centre point, especially on the surface of the sphere. And of course, the pulling force could not be equally distributed on the surface of the torn-off continent. The middle part of the continent which followed behind the ejected satellite had to travel the most and was the most vulnerable in the chain, accordingly.

An appeared kinetic energy of disintegration was the combination of two momentums: one of them was the inertia of the entire torn-off American continent with its immensely high speed towards the present west. And the second momentum resulted from the spontaneous stretching of the two American continents away from each other.

The obvious similarity between the northern coasts of the Gulf of Mexico and the northern coasts of the South American continent indicates their former unity. The correspondence of such small details as the convexity of the Mississippi River Delta with the jaw-like shape of Gulf of Venezuela resembles two pieces of an interlocking puzzle *(Pic. 15)*.

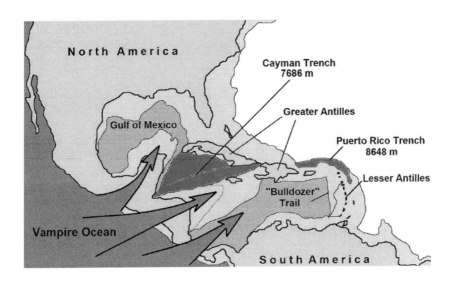

(Pic.15)

The spontaneous stretching of entire American continent led to the appearance of an enormous crack (the Cayman and Puerto Rican Trenches). Part of that bottomless crater can be partially seen under the debris of the present-day Lesser Antilles, which filled the crater.

Like in any crime scene, if there is a bullet, we must look for the gun. When we look at the Caribbean Islands, it seems like somebody tried to cover up the place. Luckily, the camouflage was not completed, and the deepest point in the Puerto Rico Trench, which is 8,648 metres below present sea level, has stayed unburied.

The stretching of the continents led to the disintegration of the 4,000-5,000-metre-high mountain range, which was the former bearing wall of our "Vampire Ocean." The Central American isthmus used to be the place of the highest mountain ranges, which was retaining the pressure of the enormous ocean. The tremendous mass of water that gushed out swept away a 1,000 kilometre wide part of the mountainous wall, filling up a newly appeared crater with debris to hide its existence once and for all.

When we look at the bottom of the Caribbean Sea, we notice a very intriguing feature there. Some enormous "bulldozer blade" left a square mark there, creating the chain of the Lesser Antilles Islands. Strange, isn't it? The flattened pile is there, but the "bulldozer" is gone. Here we can assert that **the only thing capable to plow such immense masses of earth and then disappear is ice.**

Along with debris, the water that gushed down was carrying numerous icebergs. One of the monstrous icebergs made its way through the Central American passage and slammed down into the pile, plowing everything ahead of it. Being lighter than water, it eventually resurfaced and later was carried away from the "crime scene." According to that we can surely assert that the Yucatan peninsula and also the Islands of the Greater and Lesser Antilles are nothing more than massive mounds formed by the gushing waters of the "Vampire Ocean."

The separation and expansion of the American continent also led to the breakthrough between South America and the Antarctic continents. And the name of that enormous gap is the Drake Passage *(Pic. 16)*.

The origin of this 1,000-kilometre gap differs from the Caribbean scenario. Here we can see an obvious trace of waters, which crushed and pushed the mountains 2,500 kilometres away. This "crocodile head" looking shape of the ocean's bottom is nothing more than an enormous mound of rock and earth plowed away by the gushing waters of a high-altitude ocean.

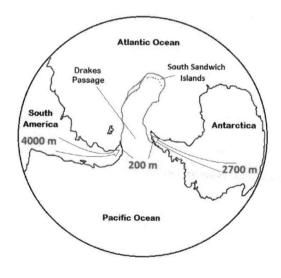

(Pic. 16)

The breakthrough of that part of the wall was not just a local event. The entire South American and Antarctic continents inclined towards the passage. The altitude near the Lake Titicaca highlands is around 4,000 metres above present sea level, which gradually decreases towards the passage. An identical picture is on the Antarctic continent. The mountain height decreases from a 2,700-metre altitude to roughly 200 metres above the present sea level.

Here I must mention the fact that, unlike in mathematics where the greater the number, the higher the value, a 4,000-metre altitude in the proximity of Lake Titicaca can be less or equal to 2,700 metres high of Transantarctic Mountains. First, the weight of the ice pushes the Antarctic continent down. The second and more important aspect here is latitude. Lake Titicaca is located around 8,200 kilometres away from the South Pole and around 2,800 kilometres away from Earth's equator, and basically 2,800 kilometres away from the highest point of the equatorial bulge. The Transantarctic Mountains, on the other hand, are located just 500 kilometres away from the geographical pole, and if we imagine Earth's complete stop, the shrinking of the equatorial bulge and the ascending of poles will show a completely different ratio.

So, at the time when the two continents separated from each other, waters of the "Vampire Ocean" gushed through the Drake Passage. The mountains, which were still holding the pressure of water, underwent an immense washing off of their coasts, which led to the collapse and slanting of the entire South American and Antarctic continents towards each other.

Evidence for that statement can be found in the scientific works of the great researcher Hans Schindler Bellamy (1901-1982). I do not want to retell his scientific observations, so here is the complete citation from his book *Built Before the Flood*.

> *It has been carefully surveyed for a length of about 375 miles. And then it was established— that it is not 'straight'. It was found that the Inter-Andean Sea of the Intermediate Level was not merely a Lake Titicaca of higher level extending far to the south, but that its level showed a slant of a most peculiar character in relation to the present ocean-level, or, what amounts to the same, relative to the present level of Lake Titicaca.*

The level of the Inter-Andean Sea revealed by the ancient intermediate strandline was higher to the north of Tiahuanaco and lower to the south.

The actuality of this peculiarity cannot be doubted, for it was established independently by different persons at different times, using different methods of surveying.

The northernmost point at which the former strandline of the Inter-Andean Sea of the Intermediate Level has been surveyed is on the mountain-slopes near Sillustani, to the west Lake Umayo in the Peruvian Department of Puno. There the former littoral is about 295 feet above the present level of Lake Titicaca, whose surface is 12,506 feet above sea-level. At Tiahuanaco, at the southern end of Lake Titicaca, the same strandline is 90 feet above the level of that great sheet of water, and 4 feet below the coping-stones of the parapets of the long-dry harbors and docks and canals of that mysterious metropolis. The ancient strandline and the ruined prehistoric city are linked beyond any doubt. The height of the strandline relative to the ocean-level decreases the further south we go. At the northern end of Lake Poopo, on the mountain-slopes south of Oruro, it is 12,232 feet above sea-level, or 181 feet above the level of Lake Poopo, or 274 feet below the level of Lake Titicaca, or 364 feet below the level of the same ancient strandline in the latitude of Tiahuanaco. Still further to the south, it is discernible just a few feet above the level of Lake Coipasa. It becomes lost in the Salt Desert of Uyuni, some 12,000 feet above sea-level.

From Sillustani to beyond Lake Coipasa, a distance of about 375 miles, the strandline dips about 800 feet. A peculiarity of this dip is that it seems to be progressive. In the first quarter of the distance it is only about a foot and a quarter per mile, while in the last fourth it increases to more than two feet per mile. This phenomenon is not without significance and yields a prop for the theory of a girdle-tide. (Pages 54-55)

On the chart drawn by Hans Bellamy *(Pic. 17)*, we can see the slant on a more visualized picture.

Hans Bellamy used the term "Inter-Andean Sea" to address the high-altitude sea occupying the area from the present-day Lake Titicaca to the desert Salar de Uyuni. According to his observations, the highest point of slanted strandline starts at an altitude of 13,550 feet, which is 4,130 metres. The present altitude of the Tiahuanaco site is 3,863 metres, which is 267 metres below the maximum level of the "Inter-Andean Sea" at that time. The sudden slanting of the entire continent led the waters of the high-altitude lake to flow towards a newly created Drake Passage. The fact that some structural remains at Tiahuanaco site are still intact is quite miraculous. Along with the sudden 267 metre drop, it was flattened by a massive volume of out-flowing water.

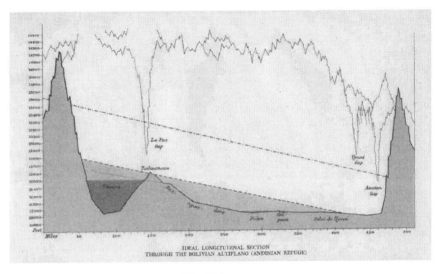

(Pic.17)

Here I must repeat myself: according to many scientists, the archaeological site of Tiahuanaco constructed from monolith blocks was built as a seaport where one of the wharves could accommodate hundreds of ships. This discovery leads us to two completely opposite assertions:

1.   Building seaports on shores of a relatively small lake to accommodate hundreds of ships would be rational only if it had access to the open sea. If that is the case, then we can surely assert that Lake Titicaca was part of the "Vampire Ocean."
2.   On the other hand, we should not forget that wharves capable of accommodating hundreds of ships were built before "The First Great Catastrophe" and therefore must have been built by a race of giants. Any remains of impressive wharves (by size) for us could have been built to simply accommodate boats for them, which means that Lake Titicaca could have been autonomous from the "Vampire Ocean" lake.

So for future archaeological excavations in the Andes we can certainly assert that any findings below the line of separation belong to the post-ejecting epoch. Artifacts found above that line could either belong to the pre-ejection period or they are remnants of an epoch that followed the cataclysm. This altitude line is a border between the past and the present. This is a gate into the other side of time when the face of Earth along with all the living forms on it was completely different.

By tracing the aftermath of ejected satellites, we cannot miss such an obvious scar on the surface of Earth as the Mediterranean Sea. The separation and shifting of the North American continent from the European continent also pulled an immense amount of land behind itself, where multiple cracks in the crust led to the appearance of the present Mediterranean, and Black and Caspian Seas, which bear all the scars of that event. What is intriguing in the shape of Europe is its diversity of mountain ranges. Knowing the process of mountain creation, the range has to have the shape of an arch, but we do not see them in present Europe.

In search for answers we must visually push the European continent back towards Africa, and we will see very distinct similarities between the coasts (Pic. 18).

(Pic.18)

Starting from west to east, let us join the south coasts of Spain and France with the coasts of Morocco and Algeria. One of the most obvious clues is the Apennine Peninsula; it perfectly fits into the Gulf of Gabes and the Gulf of Sidra. The Balkan and Anatolian Peninsulas were joined with the north of Egypt and the Saudi Arabian Peninsula. Here I must point out that the Anatolian Peninsula which was swivelled around its axis located on the peninsula's eastern end. A larger expansion of its western part is identically reflected on its northern sea. The western side of the Black Sea, which is also a crack in the crust, is much wider than the eastern side.

The deepest point of the Mediterranean Sea is 5,267 metres. If we look at the underwater terrain of the bottom of the Aegean Sea, we can easily detect an enormous landslide which filled up the deepest crater in the sea with exactly the same handwriting as in Caribbean Sea.

Considering that all the mountain ranges were formed by rising equatorial bulges, then all of them must have a curved shape. All the mountain ranges in Europe have very strange and unnatural forms, which can lead us to an important clue. The Caucasus Mountains, for example, have gradually dissolved down along the coast of the Black Sea. The Carpathian Mountains are twisted in a zigzag shape.

If we visually eliminate the Mediterranean and Black Seas by pushing the whole European continent towards Africa, we will get a completely different landscape of antediluvian Europe. By connecting all the mountain ranges in Europe from east to west, we would get an uninterrupted mountainous range of Caucasus-Pontic Range-Balkans-Dinaric Alps-Alps-Pyrenees. The mountain ranges would get their original arch shape.

By continuing our "tour" to the present south-east we cannot miss a very important feature of our geography formed during the disintegration of the European continent from the mainland *(Pic. 19)*.

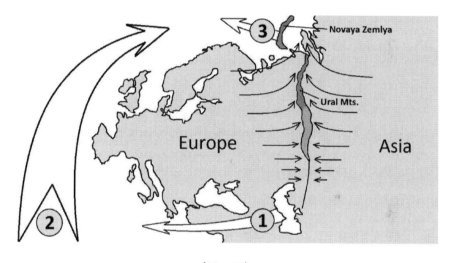

(Pic. 19)

Due to the spherical shape of the globe, a pulling and tearing force of the disintegrating part of the earth could not be directed in one direction. To trace the progression of that process we must mark three major forces of disintegration:

1. The pulling force of separation led to the appearance of an enormous crack in the crust — the present Caspian Sea.
2. A second curved force which was directed to the present north-east had swivelled the entire European continent towards Asia, with its centre point located on the eastern shores of the Caspian Sea. The Mountains of Ural, including the island of Novaya Zemlya are just piled up against the crusts of Europe and Asia.

The person who made a division of Europe and Asia by the line of the Ural Mountains was absolutely right because that is the collapse line of the swiveled European and motionless Asian "tectonic plates." Elevated at much higher altitudes, the northern parts of the Ural range indicate a much greater force on the north rather than on its south. The gradual elevation of the Ural Mountains from south to north proves the plausibility of this theory. As we see in the example of the Ural Mountains, the so-called "tectonic plates" cannot overlap one another; they simply wrinkle. The applied force had pushed the crust upwards and not under one another.

3. After the formation of the Ural Mountains, including the range of Novaya Zemlya, the force directed to the west just tore the range off from the continent creating the island of Novaya Zemlya, which is in fact the continuation of the Ural Mountains.

A topographic map of Central Asia also has many stories to tell *(Pic. 20)*. Seemingly, different mountain ranges of Caucasus and Kopet Dag mountains are in fact parts of the same mountainous system. The force that created the Caspian fracture was directed in parallel to the range and therefore divided the formerly united mountain range in two. The separation of the second satellite created a pulling force towards the centre of its ejection. The mountainous ranges of the Iranian Plateau elongated towards the centre of the Indian Ocean, and the shape of Hindustan Peninsula itself indicates an extensive stretch of the crust.

As mentioned before, the ejection of satellites was accompanied by the ejection of countless debris scattered around the globe. One of the best-preserved traces of their bombardment is located on the Asian continent. A topographic view of the area around present-day Lake Balkhash tells a conclusive story. The round mound located at its north indicates traces of an enormous wave. At some point in Earth's history, a high-altitude lake, much greater than the present Lake Balkhash, had broken its borders and carried all this debris towards the present north, partially covering the "Valley of Ponds" (as I named it).

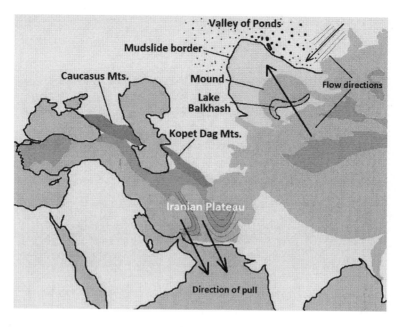

(Pic. 20)

Precious information we can read on this map is the chronological chain of events. After the ejection of satellites, the entire planet was bombarded by debris falling back from the disintegration, and myriads of round shaped ponds on the Asian continent are their silent proofs. The stretching and displacement of the entire crust led to the disintegration of many high-altitude lakes. Enormous masses of water trapped among mountainous ranges acquired escape channels and were drained down rapidly. One of the best examples of this process was the disintegration of the former "Great Balkhash Lake," where water poured towards lower altitudes had carried all the soil and debris towards the bombarded and already craterous plain of Central Asia. Pits filled with soil were buried forever and after gradual evaporation, craters filled with fresh water remained as myriads of round lakes.

The inclination of the entire South American continent is not unique for that region of our planet and an opposite coast of the former "Vampire Ocean" bears another trace of those catastrophic events *(Pic. 21)*.

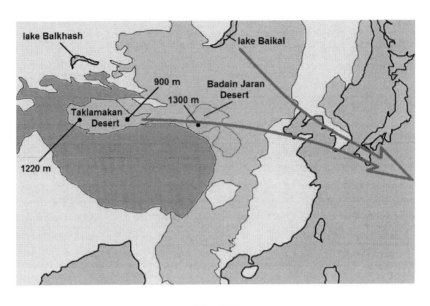

(Pic. 21)

The best indicators of continental inclinations are basins of former lakes. Deserts trapped between highland mountains are silent remnants of former lakes where water and water only can smoothen the bottom of the basin to create an almost perfectly levelled surface. The seemingly horizontal surface of highland Desert of Taklamakan shows a completely different picture *(Pic. 22)*. A virtual flood simulation of the area indicates that for water to reach the southern shores of former lake (1,550 m above the present sea level), the entire continent would have to be under water. Such a drastic inclination of the entire Asian plateau towards the present east was caused by a tremendous pull towards the centre of the Pacific.

A force directed towards the Pacific Ocean had pulled the entire Asian continent to the present east, tearing off the Korean Peninsula and Japanese Islands from the continent. According to the present landscape of the Asian continent, we can surely assert that the tension of the pull was applied not only on coastal regions. Lake Baikal, located 2,000 kilometres away from the coast, is the result of such a tremendous pull of Earth's stratum.

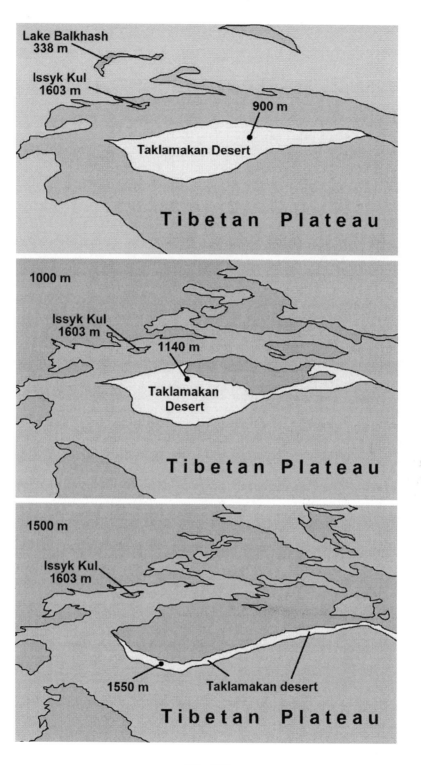

(Pic. 22)

Considering the enormous impact of the "Vampire Ocean" on every living being that used to live on its coasts, we must try to link all the civilizations of that pre-catastrophic epoch. Knowing the purpose of the megalithic walls in South America, we cannot dismiss the possibility of the existence of similar prehistoric structures around the globe.

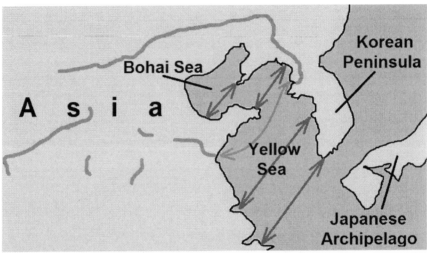

Part of present Great Wall of China

(Pic. 23)

Let us look at the map of the Great Wall of China *(Pic. 23)*. The grandiosity of such a structure has always fascinated us by its scale and purpose, but let us forget for a moment about the methods and purpose of its construction and concentrate on its location. What is most intriguing is the strange direction of the wall, where two of its sections end at the Yellow Sea. Considering that the present Korean Peninsula was torn away from the Asian continent, their virtual merge joins the wall into one. Sections of the Great Wall interrupted at the sea clearly indicate their originally united and completely non-military purpose.

The discovery of the wall by our far ancestors, its reconstruction and following re-usage in military purposes misled our present civilization of its original purpose. Similarly, with the megalithic walls of Saksaywaman and Ollantaytambo in South America, the prehistoric section of the "Great Wall of China" was serving as a dam to hold the waters of the same "Vampire Ocean."

(Pic. 24)

By traversing farther to the south-east of the continent of Australia *(Pic. 24),* we notice that the present seemingly roundish shape of the continent is the result of the splitting of its mountainous regions in two. The apparent crack between them was rapidly filled with the inundating waters of the "Vampire Ocean," along with debris carried with it. Here we can see that the mountain range on the east of the continent perfectly fits around the highlands of the continent's west and an enormous gap between them at present is just plain filled with debris. After rapid "plastering" of the area between two highlands and following the retreat of the world waters, we now observe only one unified continent.

Frozen in time

From the first discoveries of the frozen-to-death animals of Alaska and Siberia, we are always in search of answers for such a rapid freezing of all the living things that were buried alive in everlasting ice. Knowing the exact chronology of the tragic events which took place on Earth thousands of years ago, the cause of their deaths becomes easily explainable.

Let us start with our present North Pole. Do not forget that the Arctic Ocean located there is also the result of an enormous crack that appeared because of the ejection of the first satellite, and after the opening of the Caribbean and Drake Passages, waters of the "Vampire Ocean" gushed out towards the lowlands of the planet, including the basin of the present Arctic Ocean.

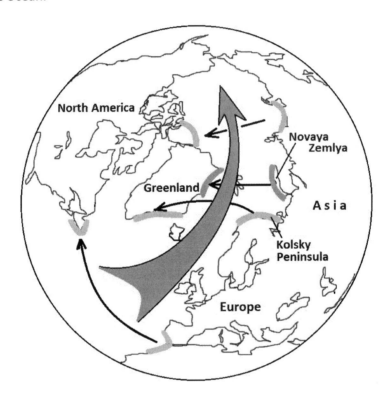

(Pic. 25)

Considering that during the reign of the "Vampire Ocean," the land pole of the greatest Ice Age on Earth was in present-day Western Europe then broken through the Caribbean Passage, the waters of the "Vampire Ocean" had to gush through the former pole towards the present-day Bering Strait. By filling the basin of the recently opened abyss, constantly upcoming waters began to inundate nearby continents by smashing everything in their way and carrying all the debris deeper and deeper to dry land.

As we can see on the map above *(Pic. 25)*, the shape of the island of Novaya Zemlya is quite similar to the coast of Greenland, that faces it. The "Tail of Jumping Tiger" (Kolsky Peninsula) also has a striking similarity with the south coast of Greenland. But there is one more curious fact. Splitting of Earth's crust was not just local (for the point of separation) event. Baikal Lake, for example, is a result of two oppositely directed tension forces of crust. Located 2500 kilometres away from the Arctic Ocean, it was filled with the fresh waters of the former pole. Such oceanic creatures as seals were carried and dumped into Lake Baikal by the waters of the Arctic Ocean. Seals of the Caspian Sea, on the other hand, were carried there through the Mediterranean passage. Considering the landlocked position of the "Vampire Ocean" and its feeding source (mainly from precipitation), we can make an assertion that the salinity of its waters was not so high and the inundation of the rest of the planet was carried by waters having diverse levels of salinity.

Waters coming from the present north were carrying less salinity in them than waters from the Mediterranean direction. The passing of the waters through ice of the former pole also lowered their salinity. The inundation of the present Mediterranean basin was done by waters of the present Sahara Desert. Due to extreme century-long droughts in the present Sahara Desert, the waters of that rapidly evaporating sea (ocean) were also extremely saline. That is the reason for such different levels of salinity between Lake Baikal and the Caspian Sea.

As many Siberian and Alaskan sites indicate, most prehistoric animal corpses have been found in upright positions. Obviously, all drowning animals were trying to keep their heads above the water in order to breathe.

Studies of frozen corpses found in Alaska and the Siberian tundra indicate that the cause of the death of animals was not drowning only. Most of them just suffocated. The disintegration of satellites from Earth led to the hailing attack of millions of fiery spheres falling down to earth. Just imagine the spontaneous activation of all the volcanoes we know today at the same time. That is why between the layers of frozen trees, animal bones, mud and other debris there are layers of volcanic ash. In this case, there are two possible causes of suffocation:

1. From poisonous gases (formation of the satellites was accompanied by enormous heat and the incineration of half of Earth's crust);
2. From rapid freezing of surrounding waters (comparable with the anaconda prey effect, after each exhalation, the weight of gradually freezing and gradually expanding ice left less and less room for inhalation).

All the corpses ever found frozen have signs of decay, which was stopped by sudden cooling. It would be logical to assume that the occurrence of such a catastrophe would prevent any sunlight from penetrating through multi-ton layers of dust for many, many months. Considering that wet objects are cooling down at a much faster rate, all animals and vegetation of the near-splitting area were destined to be frozen alive.

Let us suppose that right after the separation of satellites the entire planet plunged into a global Ice Age. After gradual warming, all the frozen carcasses of animals and trees would be defrosted and decompose. But the fact that we still unearth frozen animals in Alaska and Siberia indicates that this part of the planet was always close to the pole. In other words, after the separation of satellites, Earth was locked in the same position with minor shifts of the pole somewhere between Alaska and Siberia. Basically, due to the appearance of the moon and bitten shape of the globe, postdiluvian poles shift to very short distances.

The proof that the freezing of mammoths took place at the time of the satellite ejection lays in the global map of mammoth remains, where again, the Hudson Bay Basin is completely free of any fossils. Along with the absence of mammoth and dinosaur remains, the Hudson Bay Basin is free of oil and natural gas deposits. We are not talking about animal fossils ever found in the region; we are talking about the absence of any mass graveyards of prehistoric animals like those found in Alaska and Siberia. During the catastrophe, the enormous ice cap of the former pole was blanketing the Hudson Bay Basin and preventing it from getting any organic deposits from outside.

Previous poles located on water are untraceable, as opposed to poles located on land. Polar ice of the former pole located near the present pole, does not melt even in centuries. The snow of Greenland is the best example. The appearance of such a thick layer of snow on it indicates its previous location on a pole, and the rapid melting of its snows at present is just a recurrent process of the eventual melt down of ice shifted away from the geographical pole.

Here we can bring up a very curious epoch of Medieval Europe, the so-called Little Ice Age. It is a period of cooling in Europe and North America, which took place approximately between 13th and 19th Century AD. The Great Famine of 1315-1322 was the result of sudden cooling in Europe. Widespread crop failure led to the mass starvation and death of about 10 to 15 percent of the medieval population of Europe.

The same period was registered as the coolest in the Southern Hemisphere. Glaciations of New Zealand, and the south parts of South America and South Africa, reached their maximum extent in the mid-18th Century. The discovery of the two-thousand-year-old abandoned settlement at Point Hope in Alaska indicates that at the time of its establishment the pole was much further from it than it is now.

By visually placing the pole in Greenland and the opposite pole on the Antarctic continent, we would get an Ice Age effect in Europe. As we know, the location of both poles on dry land leads to an Ice Age even with the existence of a moon orbiting us. If we are to suggest the location of the pre-Greenland pole to be at the same spot as the present, then a sudden 2,000 kilometre shift would lead to substantial climatic changes.

It is probable that all the medieval mentions of change in the sun's direction were lost among the plague-starved and infected population of Europe. As we know, all the mass contagious diseases burst in densely populated cities and the centres of culture and science accordingly. Being undoubtedly noticed by all the people around the globe, the same phenomenon had completely different impacts. While densely populated regions were going through complete extermination, nomadic tribes had more chances to survive, but the absence of writings among them would erase any mention of that event.

Earthquakes

There is one distinctive type of tectonic fracture in the world, and it is located at the base of mountain ranges. Elevated by the equatorial bulge, mountains are the only vulnerable spots on the globe to be torn apart. In comparison to the process of rolling out dough, thanks to flour, folded layers are very hard to join together again. Similarly, thanks to the accumulation of dust and vegetation, the piling up of different strata of the Earth is hard to join together, and cavities formed by folded strata have to eventually collapse. Concluding from the above-mentioned criteria we can distinguish three major types of earthquakes on our planet:

1. Local earthquakes. Frequently occurring earthquakes in mountainous areas are due to the collapsing of cavities between compressed and piled up strata created by raised equatorial bulges.

2. Aftershocks of satellite ejections. All the earthquakes on the globe are aftershocks of the crust displacement caused by the ejection of satellites. That is the reason for the intense power of earthquakes located on the "Ring of Fire" zone and gradual diminishing power of earthquakes occurring away from it.

3. The last, science-fictional and most destructive type would be an internal earthquake, coming from the depths of the planet. Considering that our planet is a hollow sphere, collapsing of any satellite inside the sphere to the core would lead to an unimaginable wobbling of the entire globe.

Knowing the cause of the majority of earthquakes on Earth, we can easily predict locations of future strikes. Considering that the second satellite's vector of ejection was directed away from the centre of the globe then all earthquakes that occurred in the vicinity of the present Indian Ocean are aftershocks of the wave originated in the centre of its basin.

In contrast to the second satellite with its point of ejection in the centre of the basin, the first satellite was thrown away tangentially and its point of ejection therefore cannot be located in the centre of the Pacific.

Considering that the entire American continent was torn away from Asia towards the present west, then the ejection point of the first satellite should also be located west from the centre of the Pacific Ocean.

The main conclusion here is that by tracing each earthquake on Earth, we can link them to the ejection points of two major satellites separated from Earth thousands of years ago, and the movement of tectonic plates and following earthquakes are just aftershocks of that single global cataclysm.

Volcanoes

If we look at the world map of volcanoes, we can see that most of them are concentrated on the perimeter of the Pacific Ocean. That is why it is called the "Ring of Fire."

Located on the perimeter of an enormous crater left by ejected satellites, it would be more logical to consider volcanoes as Earth's wounds, but it is not the case because, in that case, all of them would be scattered on the bottom of the Pacific and Indian Oceans only. But they are not. And again, in search of answers, the best place to go is South America, where we can see that all the South American volcanoes are located on the Andes, the mountain range with an average height of 4,000 metres above the present sea level. The average depth of the Pacific Ocean is 4,000 metres below the sea level. Considering that the first satellite was torn away from the basin of the present Pacific Ocean, the approximate difference between the highest point of the Andes and the lowest point of the Pacific Ocean would have to be around 8,000 metres. If all volcanoes on Earth come out from its core, then why did they choose to break through the highest altitudes of the planet instead of finding paths shortest to the surface?

The ejection of the satellite led to the sudden disappearance of an 8,000-metre-thick stratum of Earth. Whether it is a volcano or any other liquid substance, it always looks for the shortest and weakest spots on the surface. The existence of all the South American volcanoes on top of the Andean mountains instead of being 8,000 metres closer to the source (the so-called molten core) just does not make any sense. Perhaps our entire conception of volcanoes is fundamentally wrong.

To answer that question, we must visualize the process of satellite ejection. Along with the separation of two main satellites from the surface of Earth, there were myriads of other, much smaller debris. From the size of an egg to the size of mountains, all the debris that did not attain escape velocity eventually hailed down on the Earth. From the moment of their formation to the time of their fall, the smaller objects cooled down and hailed down in the form of spheres. At the same time, the cooling down of mountain-sized objects has a completely different cooling ratio.

If Earth's core is still hot, then the volcano-forming process should be one continuous process, which means that along with already established volcanoes, there should be the constant appearance of newborn craters around the globe. But leaving all the lowlands of our globe, which are basically closer to the "boiling core," volcanoes have chosen the highland terrains. Odd, isn't it?

The answer is fairly simple. After smashing a perfectly round snowball to the wall, we get a perfect cone. It would be very hard for our scholars to accept, but all the volcanoes of our planet are just fiery debris of ejected satellites smashed back on the Earth. After centuries of cooling, the exterior crust began to harden, sealing the heat inside. A constantly boiling mixture excreted gases accumulated in its cavities creating enormous pressure. Since gases evaporate upwards, after reaching the breaking point of already cooled down rocks, they simply burst their way out into the sky. As we know, water boiling in an uncovered pot will eventually evaporate completely. A boiling pot with a sealed cap will lead to an eventual explosion. Similarly, smoking volcanoes can smoke forever and never erupt; they just gradually cool down.

The way we picture volcanoes today is just multiple pores on the surface of the planet periodically emitting molten lava from the centre of the globe. Now, knowing the structure and biography of our planet, we can certainly assert that <u>volcanoes, the way we know them</u> as the belching lava from the centre of the globe simply <u>do not exist.</u>

The locations of the majority of volcanoes on high altitudes have wrongly led us to presume that volcanoes are the main cause in the creation of mountains. But in fact, volcanoes existing on the tops of mountain ranges are just perfectly preserved remnants of those catastrophic events. Debris fallen to the ground were momentarily swept away by enormous tsunamis of the "Vampire Ocean." Volcanoes located on the bottom of the ocean, for example, cannot erupt like volcanoes located on the surface. A multi-ton mass of water constantly prevents underwater volcanoes from erupting; the water just calmly absorbs their heat. Water boiled to hundreds of degrees simply has no chance to evaporate.

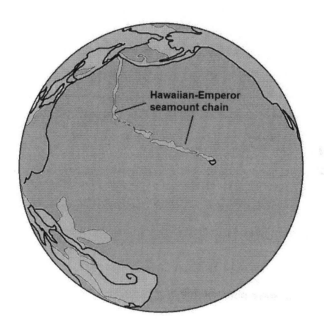

(Pic. 26)

Considering the locations of the present north and south directions, we can assume that the American continent was shifted from east to west, and the newly separated satellite was ejected tangentially towards the present west. During its separation, the newly formed satellite took with it

77

a tremendous volume of soil and water, which later hailed down to Earth. Located in the Pacific Ocean, the "Hawaiian-Emperor seamount chain" *(Pic. 26)* is in fact the birth cord of the newly born satellite or the broken chain of a ship floated away. Whatever we compare it with, the fact is that they are fallen debris of an ejected satellite.

We can certainly assert here, that the Hawaiian-Emperor seamount chain and all the volcanoes of the Pacific Ocean are fiery debris that shattered down from an ejected satellite. Due to the release of immense energy, a newly appeared enormous crater was covered by a solid layer of molten rock, and in addition to that was bombarded by fiery debris from a recently peeled away part of the Earth.

Considering the tremendous heat that was emitted during the satellite ejection, a significant part of the "Vampire Ocean" momentarily evaporated away. The remaining waters have gushed out through newly opened passages, destroying everything in their path. And I would assume that along with the enormous iceberg mentioned above, which plowed a huge amount of earth creating the Caribbean Islands, many icebergs of the former pole made their trips around the globe where on the way back they were beheading all the volcanoes whose tops were close to the surface. It seemed like a mass execution by cavalry armed with swords. And this is the exact cause of the mysterious form of Guyots scattered around the globe. Guyots are underwater volcano-type mountains with absolutely flat tops.

We know that the Atlantic Ocean is the result of an enormous crack in the Earth's crust, and the east of the American continent perfectly fits into the western shores of Africa and Europe. The shores of Greenland match with the shores of Baffin and Ellesmere islands on one side, Kola and Taymyr peninsulas on the other side, with the island of Novaya Zemlya in between them. What does not fit into the complete picture of the pre-ejection of satellites is the island of Iceland, one of the most volcanically active spots on Earth. Such a large, round area of constantly erupting volcanoes is in fact just one enormous volcano; it is a huge, fiery sphere, which was smashed back to Earth right after ejection of the first satellite.

Another proof for the nonexistence of a centralized volcano system can be found on planet Venus. Despite its scorching hot temperature of 462° C, there are no volcanoes ever discovered. The cooling down of Venus is flowing quite peacefully. Because of the fiery liquid substance of Venus, accumulated gases bubble up and eventually burst, creating enormous craters, in contrast to Earth, where the absence of an atmosphere and vegetation will always show these scars. Surely our planet underwent the same process and, after multiple reversals, craterous terrains were flattened down by the weight of polar ice, piled up by newly created mountains, or washed away by waters of the world ocean.

Obvious proof to the continuous thermo reactions are the volcanoes of Antarctica. What a strange coincidence. Most of the Antarctic volcanoes are extinct. As we mentioned above, volcano eruption occurs due to the accumulation of evaporated gases in the enclosed cavities of a mountain. When the pushing force of generated gases beneath the volcano prevails against the weight and strength of the rocks above, the volcano just wakes up. In case of the Antarctic volcanoes, a heavy blanket of ice holds the pressure of pushing up gases. Since all volcanoes have recurrent eruption cycles, during the periods between eruptions, constantly accumulating ice eventually sealed the caps. As long as a champagne cork holds the pressure, the gaseous beverage will remain restrained. Shaking the bottle and loosening its cap will lead to the sudden burst of the mixture. Identically, any earthquakes around volcanoes agitate their spontaneous eruption.

Today's Antarctica is home for many international bases. It is quite understandable to have far-sighted plans regarding our frozen continent because after the next recurrent reversal, the Antarctic continent can be located away from the pole, which means the appearance of favourable climatic conditions and the attainment of new territories. But there will be one big disappointment there. The gradual disappearance of ice will trigger spontaneous eruptions of all the Antarctic volcanoes. The immense heat from the "sleeping" volcanoes cannot just disappear; it just waits for the appropriate conditions to escape.

So, now we know that volcanoes and some islands were formed as a result of the enormous fiery debris of satellites falling back to earth. Due to their lighter weight, smaller fragments managed to cool down before hitting the ground, where such impact formed craters on completely flat terrain.

Unlike debris which hit the ground, fragments splashed into the water could preserve their spherical shape. The so-called mysterious spherical stones of Central America are just debris of an ejected satellite. Nowadays these rocks are constantly getting unearthed all over the world in the island of Spitsbergen, Bosnia, Russia, etc. Their large concentration on the western shores of Central America can be easily explained. Taking into consideration that the separation of satellites was a worldwide event, it would be hard to imagine any region of our planet not scattered with such perfectly round rocks. The only difference is the depth they are buried under. After falling back to earth, subsequent dust and water buried them underneath. Fragments that fell into the former "Vampire Ocean" were carried away through the narrow passage of the flattened Central America to fill the newly appeared abyss. After the gradual counterbalancing of waters of the World Ocean when the volume of gushing water ceased, remaining debris piled up on the western shore of a narrow strip of Central America forever.

But as we know, besides those spherical rocks, Central America is famous for its spherical, carved Olmec heads. And again, their location on that narrow strip of land is to be expected. These perfectly spherical stones were carried and left on the west coast of the isthmus by gushing waters of Vampire Ocean. The existence of such smooth spherical rocks is a precious find for any stonemason, and ideally the only spherical thing to depict would be that of a human head. Now, knowing the origin of these perfectly spherical rocks, we can certainly assert that the Olmec stone heads belong to a civilization flourishing in the region after that terrible event.

Another mysterious creation of our planet is the honeycomb shaped rock formations scattered around the globe *(Pic. 27)*.

As we know from the laws of physics, the sudden cooling of any substance leads to its crystallization, and the typical shape for crystallization is a hexagon. So whether it be the Giant's Causeway of Ireland, the Devils Tower of Wyoming, Sawn Rocks in Australia, the hexagonal-shaped rocks of Anatolia, or any other regions of Earth, all of them are remnants of rapidly cooled down debris of satellites that have fallen to Earth.

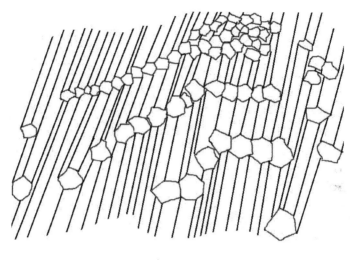

(Pic. 27)

The existence of these hexagonal rocks in particular regions of our planet gives us very important clues about the locations and extension of polar ice of the pre-satellite ejection period because rapid crystallization of overheated magma would only occur due to its sudden interaction with ice. Considering that, before satellite ejection, the North American and Eurasian continents were parts of one united continent, then we can surely assert that two impact sites, the Giant's Causeway of Ireland and the Devils Tower of Wyoming, were covered by the ice of the same Hudson Bay Basin pole.

By tracing the debris of the ejected satellite, the North American continent has a complex story to tell. First, we have to mention that right before the cataclysmic ejection of satellites, the North American continent was the bearer of one of the poles. A spontaneous shift of the continent and the massive weight of polar ice caused the appearance of multiple cracks in the crust. The Canadian Arctic Archipelago and the Great Lakes of

North America are the result of the immense stretching of Earth's crust. To explain the complex geography of the Rocky Mountains *(Pic. 28)*, we must try to reconstruct the condition of that region prior to its disintegration from the Eurasian continent.

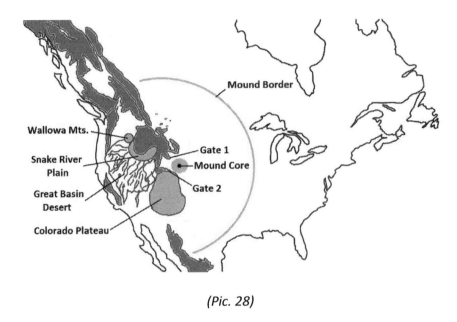

(Pic. 28)

The first place we must start is by pinpointing the location of the pole in the Hudson Bay Basin. Of course, we can only guess the scale and extent of that former polar ice, but thanks to the present South Pole located on dry land, we can make a conclusive comparison:

1. Located on the Antarctic continent, the present South Pole and former pole of Hudson Bay have one distinctive similarity: both of them are (were) situated on dry land.

2. If present-day Earth (which experiences a change of seasons) accumulates so much snow in Antarctica, then the absence of tilt of rotation would lead to much more severe climate and widespread cover of ice in polar regions.

3. The entire Antarctic continent at present is covered by snow, including the very tip of the Antarctic Peninsula, which is located almost 3,000 kilometres away from the pole.

4. From the above-mentioned conditions, we can make an assertion that the ice sheet of the former pole (the present-day

Hudson Bay Basin) could easily reach the shores of the "Vampire Ocean" (the present-day Pacific Ocean) located the same 3,000 kilometres away.

The topography of the Rocky Mountains at first sight does not look any different from the rest of the globe except the "belly" of the Great Basin Desert, the largest high-altitude desert in North America. But besides that enormous valley of salt, there are many other clues to be detected. The perfect half-round shape of the plateau gradually descending towards the Hudson Bay Basin *(Mound Border, Pic. 28)* was achieved thanks to the smooth layer of ice. Debris that gushed through the "gates" was equally spread over the smooth surface of ice and formed a perfectly round mound. The mound border in fact is the border between saline debris and uncontaminated parts of the ice cap; it is the border between the prairies and fresh water lakes of present-day North America.

The next obvious arch-shaped landscape is the Snake River Plain, which is pointing at the enormous wavy plateau of the Great Basin Desert.

1.   The only thing capable of leaving such a semi-round imprint is a wave of large debris impacted to ground.

2.   Pointed away from the arch, the wavy landscape of the Great Basin Desert indicates that an enormous chunk of earth fell into the water.

3.   The high concentration of salt in the desert can indicate two things:

   a)   A chunk from the ejecting satellite fell into the saline lake.

   b)   The meteorite itself was an enormous "drop" of ocean water lifted up by the ejecting satellite.

Considering that the "Vampire Ocean" was not extremely saline, the most probable scenario here would be option (a). Considering the enormous size of fallen debris, the volcanic activity of such a massive chunk should continuously erupt during several millennia and in tremendous proportions. But it does not happen, and there are no traces of any magma flow. What we have there instead are just numerous geysers. But geysers are extremely hot because of their contact with immensely hot magma.

Considering that the enormous fiery mixture impacted the ice sheet of the previous pole, an entire layer of high-altitude ice was just sandwiched between the ground and the molten lava above. Water, having nowhere to escape, periodically bursts into the air to flow down again. The existence of geysers means the considerable stability of the hot reactor beneath. The gradual power loss of these geysers can mean two things: the shortage of circulating water, or the magma had finally cooled down. The second scenario would be much more preferable because the absence of a coolant will eventually bring it to explosion.

Considering that during the separation of the first satellite, the rolling up of the crust was directed away from the pole located on dry land (the Hudson Bay Basin), it would receive most of the largest debris. Like in the case of a car stuck in the mud, skidding tires would throw the debris in the opposite direction. Here we can detect the exact same picture: the largest debris from the satellite ejection is in North America and the overlapping strata of the last layer of the rolling up satellite is located on the opposite side of the Pacific (Fiji Basin and Mariana Trough on the bottom of Pacific Ocean, *Pic. 29)*

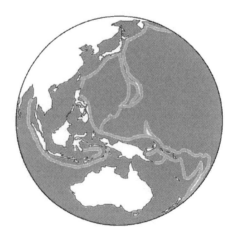

(Pic. 29)

The satellite image of the Salmon River Mountains of Idaho and the Rocky Mountains of Montana shows a distinctive oval shape of a 480-kilometre-long region, which looks like an enormous bug with a "camouflaging eye" on its back that landed in the middle of a small pond *(Pic. 30)*. The impact of such magnitude created immense waves close to the shore regions of the "Vampire Ocean." According to the direction and size of the dunes in the Great Basin Desert, we can imagine the force of impact.

Normally, an object dropped into a unified substance creates perfectly round waves directed away from the point of impact. Taking into consideration that our "bug" is located right in the middle of a mountain range, we can try to put together the scheme of impact.

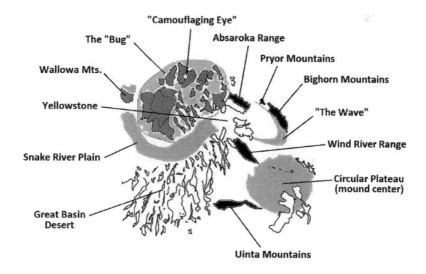

(Pic. 30)

The Rocky Mountains of North America back then were located right between two powerful forces of nature: the constantly raising waters of the "Vampire Ocean" on the west and the constantly spreading polar ice on the east. At the time of ejection, both natures have reached their maximum magnitude. Fiery chunks of satellite debris were falling into the water, creating immense splashes of water followed by enormous waves. Meanwhile, the part which hit the frozen mountain range was met with greater resistance. It does not mean that the kinetic energy of the fallen "meteorite" just disappeared into nowhere. The waves appeared after such a massive impact tore and pushed the entire mountain ranges away.

Carried on top of the wave, the Bighorn Mountains of north Wyoming separated from the Absaroka Range, leaving its disintegrated part behind (the Pryor Mountains). The "monster" that crushed to the ground had also opened 180-kilometre-wide "gates" to the east. Pushed to the side, the Uinta Mountains and the Wind River Range are just silent remnants of a suddenly opened enormous passage.

Here it is necessary to mention that at the time of those events, vast glaciations covered the entire North American continent, where all the roughness of the landscape was smoothed-out by ice. Debris that gushed onto the flat surface got evenly distributed in every direction forming a perfect semi-circular shape. After the following polar shift and gradual melting of the polar ice, the giant mound eventually sank down to the ground, pressing and trapping the water beneath. That is the exact reason for such a perfect cone-shape and circular borderline of the North American prairies. By placing its centre right between the two opened gates we get approximately 1,500 kilometres to the north and south. Its shorter expansion to the east (roughly 1,200 km) can be explained by its nearness to the pole where a higher level of snow played the role of barricade and did not let the landslide to spread any farther *(Pic. 28)*.

The other unique feature of the North American continent is the Colorado Plateau. Famous for its distinctive bright-orange colour, Red Rock Country seems fundamentally different from the rest of the continent. The only similarity to the Colorado Plateau can be found on the other side of the Atlantic: the Great Sahara Desert. Knowing the history of the Atlantic Ocean, we can surely assert that right before the continental fracture the Great Sahara Desert and the Colorado Plateau were part of the same ocean basin. The orange coloration typical for these two deserts is due to their exposure to extremely high temperatures for an extended period of time. Such a "fried" surface of the deserts can be achieved only during extremely hot and single seasoned years on the equator and the Ice Age period on the polar regions of the globe accordingly.

The best observational cross section of Earth's crust is in the North American Grand Canyon plateau, which was rapidly pulled towards the ejection and then cracked during its following retreat after the ejection. We experience the same effect every day. During the abrupt slowing down of a vehicle, all its mass leans forward and after its complete stop, it leans back again. Identically, the entire Colorado Plateau was pulled towards the ejecting satellite and after complete separation of the satellite, its abrupt retreat formed an enormous crack.

This unique (<u>not carved by the Colorado River</u>) plateau is the bearer of precious information of its formation. In 1869 during his expedition to the Grand Canyon, the great American geologist and explorer, John Wesley Powell, named the border between the two sedimentary strata of completely different epochs of the Grand Canyon as the "Great Unconformity" *(Pic. 31)*.

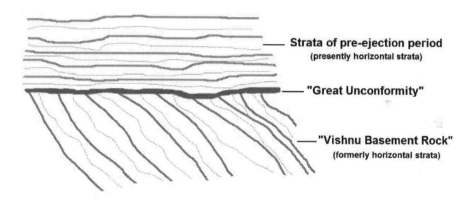

Strata of pre-ejection period
(presently horizontal strata)

"Great Unconformity"

"Vishnu Basement Rock"
(formerly horizontal strata)

(Pic. 31)

The metaphoric rock of the lowest part of the Grand Canyon completely differs from all the multicoloured layers above, and that fact has puzzled many generations of scientists. In fact this very "unusual" basement layer of the Grand Canyon, called the Vishnu Basement Rocks, extends across the entire North American continent from the east coast to the base of the Rocky Mountains, and from Greenland to the Gulf of Mexico.

The undoubtedly scientifically valuable crosscut of Earth's crust at the Grand Canyon shows the following:

1.   The tilted plateau of the Vishnu Basement Rocks indicates only one thing: it was raised from the bottom of the prehistoric ocean during the ascension of a newly emerged equatorial bulge.

2.   The existence of such a thick horizontal layer indicates that it was submerged under water for quite an extended time, and it was formed way before the appearance of the "Vampire Ocean."

3.   Such numerous quantities of diverse layers of sand indicate different epochs of this formerly submerged region, where every stratum differs from neighbouring layers, and belongs to one particular epoch with its depth below the surface of water and its location on the globe.

Here we can make the following conclusion: during many epochs before the appearance of the "Vampire Ocean," as part of the ocean bed, the Grand Canyon and the Great Sahara Desert experienced multiple changes to their locations and climates accordingly. By exploring the presence or complete absence of fossils in each particular layer, we can determine the location of this region with respect to the poles at each particular epoch.

Heading further north-east we can detect many traces of heavenly "droplets." The enormous force of a rolling up satellite was spreading its largest debris to the direction opposite to the separation. The semi-round coasts of Hudson Bay and the Gulf of St. Lawrence are obvious indicators of such bombardments *(Pic. 32)*.

Here again, we must mention that during the existence of the "Vampire Ocean," the present North American continent was covered by a layer of ice enormous by size and thickness. Any fiery bombardment of former polar regions would meet enormous resistance of centuries-old sheets of ice, and that is why enormous shockwaves of "droplets" had indented the ground beneath, creating round-shaped imprints. The absence of impacted material indicates that the remains of the fallen debris were carried away by the melting waters of the former polar regions. Besides the above-mentioned craters, we can observe numerous round-shaped lakes in Quebec and Labrador where all of them — without exception — are remnants of the catastrophic events of the satellite ejection.

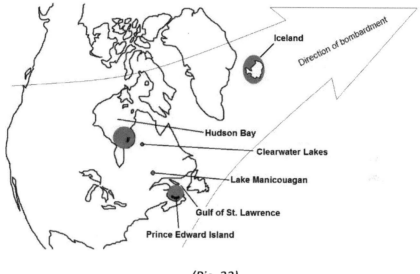

(Pic. 32)

The next large footprint of heavenly bombardment is in the waters of the Atlantic Ocean. Thanks to the development of satellite imaging we can clearly see that the underwater landscape of the Atlantic Ocean is covered by numerous mounds scattered by the ejected satellite. One of the largest and tallest among them is the island of Iceland. Its altitude above sea level makes it the most volcanically active place on Earth. Thanks to its enormous size, it remains elevated above the sea level. Mounds located below the sea level were cooled down by freezing waters of the polar regions. Of course, it does not mean that all the underwater volcanoes are now extinct. An enormous weight of water holds the pressure of the "champagne cork" and prevents it from erupting. After gradual, centuries-long cooling, the force pushing up the heated mixture will eventually diminish.

By tracing the trajectory of the ejected satellite, we get to the present European continent, which also bears many scars of the fiery bombardment. Satellite images of the Mediterranean region show distinct traces of impact where arch-shaped forms of the Cantabrian Range and Central System on the Iberian Peninsula in fact are the outskirts of one large impact crater *(Pic. 33)*.

The crater in central Europe surrounded by the mountains of Sudety, Ore and Bohemian Forest is an enormous dent left by debris impacted to the ground, which sliced the mountain range in two, separating the Carpathian Mountains from the Dinaric Alps. We must note that the impact crater is located right in front of the spontaneously opened "gate" of the Carpathian Mountains. Of course, any impact directed along the mountain range would have a much greater destructive effect than if it were directed at its front. In other words, to cut a piece of wood lengthways is much easier than to break it in two. That is the exact reason for the zigzag shape of the Carpathian Mountains.

A similar picture can be found on the north of the African continent. The Mountains of Tibesti and the arch-shaped hills on its west are the result of the wave of impact. Thanks to the appearance of topographic maps, it became possible to detect all these distinctive traces of catastrophic events of the past.

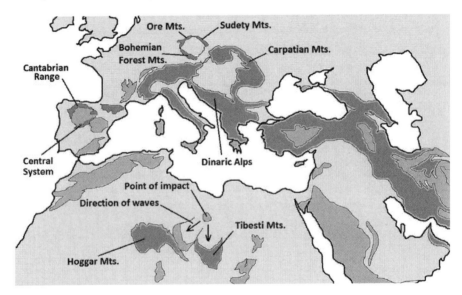

(Pic. 33)

In conclusion, the causes of mountain formations on our planet can be classified into four main groups:

1. Mountain ranges formed by newly developing equatorial bulges in the <u>absence</u> of any satellite around the globe. (Due to their nature of piling-up around the heavy "blanket" of ice). (Himalayas, Andes, etc.)
2. After the disintegration of continents due to the ejection of satellites, the bombardment of satellite debris that followed caused the appearance of all the volcanoes, single standing mountains, hills and lowland craters.
3. Mountain ranges created by the wave from impacted debris. (Wind River Range, Bighorn Mts., Carpathian Mts., etc.)
4. Mountain Ranges created by the wrinkling of Earth's crust. (Mountains of Ural)

The Moon

"In the earliest times, when the

Moon was not yet in the heavens"

The Tribesmen of Chibchas.

There are many accounts around the globe of people who used to live on Earth before the existence of the Moon. According to Aristotle, Plutarch and other Greek historians, Arcadia in Greece was populated by Pelasgians, who were later called Proselenes (before-mooners).

"Tribe of Indians met in Guiana, by Humboldt about 1820, said their ancestors existed before the Moon". (Harold T. Wilkins "Mysteries of Ancient South America" p.86)

I find it strange that we have very limited information about our ancestors who used to live just 5,000 years ago, and yet the fact that some time in history (according to science it was four and half billion years ago) the earth was moonless, we remember it very well. Complete absence of any logic.

A present, the common explanation of the moon's origin is that it was captured by Earth's gravitational pull. The moon and the majority of the satellites in our solar system have circular orbits. If you recall from childhood, in order to ride the carousel, we would wait for its complete stop, wouldn't we? Of course, we were trying to jump on it on the go, but every one of us knows what happened when we did. Why do we seem to think that planets obey different laws of physics than we do?

When we meet a family consisting of grandparents, parents and children, we know the children were born to their parents, who were born to their grandparents and so on. But we never think that all of them have gathered together from who-knows-where. The only case of appearance of such thoughts would be the inability to determine the age of every member of the group.

We know that our planet Earth spins around the sun, which spins around the centre of the galaxy, which spins around the centre of the universe. And suddenly, out of nowhere comes our satellite moon and from the first attempt gets captured by Earth's gravitational pull (what nonsense). First of all, to be captured by a celestial body with such a complex trajectory and without a second attempt would require an enormous gravitational pull. And secondly, where are all those satellites coming from?

If we are to accept such a scenario when satellites are formed from their mother planets, then a newly formed satellite is nothing other than a molten sphere heated to immense temperatures, where the centrifugal force of the newborn balloon creates a fiery wave inside the satellite itself (Pic. 34). The crust of the molten sphere spinning in a weightless environment cools down way before the wave traversing inside of it.

Our ordinary mind would imagine a mass ejected from the planet as some irregularly shaped object. But in a weightless environment, whether it's a drop of water or any other liquid substance, the only shape for it is a sphere. Heat released during a satellite ejection can reach millions of degrees, converting the hardest stone we know into a liquefied substance.

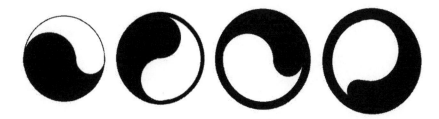

(Pic. 34)

When the glassblower blows the glass, he constantly spins it in order to get an even distribution of the molten glass inside. If the craftsman misses the moment, he can always heat the glass again and try to fix the problem. In the case of satellites, there is no second chance. Rapidly spinning in space, a "newborn" satellite forms a perfect sphere, where the wave of molten lava is travelling on the interior walls of the satellite gradually thickening its walls.

After the gradual cooling down, the internal wave of lava gets thicker and thicker until it solidifies on the interior wall of the satellite, forming a big chunk of highly compressed rock. As we know from childhood, playing with an evenly inflated ball is much more fun than playing with a ball with small amounts of water or any other heavy object inside. Unevenly distributed weight inside the sphere brings it to a complete stop. The nipple of an inflated balloon always faces down; it is as simple as that. The same way that a ball stops with its weight on the bottom, the satellites stop spinning with the frozen lava on the bottom also. The phrase "on the bottom" means "towards the gravitational pull of the mother planet." That is the exact reason why all the satellites are locked in one position towards their parental planets.

We always only see one side of our moon, which is in fact its heaviest part. And it is directed towards the gravitational pull of the earth. Gravity indicators on the moon or any other satellite cannot be due to different thicknesses of its walls. And that is why our moon has different gravitational pulls throughout its surface. Standing on the top of a massive bulge will detect maximal gravity as opposed to being located on the thinnest layer of the satellite. A larger and denser substance means stronger gravity.

According to the picture above, the size of the satellites depends on three major criteria:

1. The mass of the ejected satellite
2. The speed of ejection
3. The angle of ejection, which is the most important factor (A tangentially-ejected body has a much higher speed of rotation than the object ejected at the right angle.)

Two satellites identical in size do not necessarily have the same mass. The heaviest one can be created by slow but massive ejection. And the lightest one could be the result of an immense speed of ejection.

Having such vast basins on the surface of Earth, we can determine the size of ejected satellites and their whereabouts. To find them in space we need to have an idea of their sizes. Summarizing the approximate widths of the Pacific and Atlantic Oceans, we can obtain the size of our "Vampire Ocean." Its length would be equal to 21,800 kilometres *(sum of distances A to B and C to D (Pic. 4)*, which is slightly greater than half of Earth's diameter, the factor which played a major role in satellite ejection.

By traversing to the other side of our planet, we can see that besides the basin of the former "Vampire Ocean" there is also another quite large crater on the surface of Earth, and that would be the Indian Ocean. By reconstructing the crater of the Indian Ocean *(Pic. 2)*, we can see that it had quite an enclosed area, which could have been covered by land. Considering that the "Vampire Ocean" pumped all the equatorial waters into itself, we can certainly assert that right before the cataclysm, a vast area of the present Indian Ocean was most likely a dry land with minor water basins in it.

During the recurrent reversal, when Earth experienced a short-term disruption of its smooth rotation, the immense mass of floating polar caps had created an enormous wave with enough momentum to pull a substantial surface area of Earth's crust behind it. Attained inertia and pressure of the continent following behind it caused the peeling and ejection of more than half of Earth's crust into the space.

In a matter of minutes, the body that separated from Earth began to boil with bursting bubbles like a huge steel boiler, where molten waves were travelling around it with great speed to mould a perfect sphere. Kinetic energy of separation was so immense that it pushed this molten chunk of Earth away from its mother-planet with great speed.

The separation of such an enormous section of Earth's crust dragged behind a massive layer of the planet, peeling more than a half of Earth's surface. Along with the instantaneous displacement of the entire American continent, a shower of countless igneous spheres followed behind the main stirrer of this destruction. Part of them flew out to space, and others hailed down to Earth in the form of deadly fireballs carrying more and more destruction to an already wounded planet. Like giant catapults stuffed with flaming bullets, they were bombarding the earth.

The spontaneous shift of such a tremendous continental mass (more than half of the Earth's surface) towards the point of ejection had created an immense tension directed away from the point of launch. Considering that the satellite's tangential ejection was directed towards the Asian continent, the force pulling it back was directed towards the present-day east. As in the case of stretched rubber, the tension force of Earth's crust eventually pulled the most outstretched continents back. The wave that appeared as a result collided on the opposite side of the planet causing the ejection of another, much smaller chunk of the crust into space. Minutes after the first launch, the second satellite took off into space carrying another quarter of the original surface of Earth. And in the same way, this fiery, red boiling and bubbling sphere was rolling and flipping around to form a perfect sphere again.

But this time the inertia of the ejected body wasn't directed tangentially at the Earth's surface. It was directed at the right angle and had much less inertia, which made this fiery ball stay in Earth's orbit. Basically, the second "baby" was born because of the stretching out of all the continents encircling the globe in addition to the massive impact from the separation of the first sibling. The birth of all these siblings was accompanied with enormous waves of water, which were chaotically splashing around the overturned planet, irritating its already open wounds.

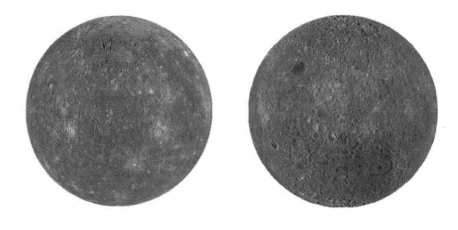

(Pic. 35)

Now, having a complete picture of that event, the only remaining task is to find those satellites. Let us look at the picture above *(Pic. 35)*. The satellite on the right is the back side of our Moon. We can clearly see all the round hills and countless craters. The ideally spherical shape of the satellite was achieved by it creating immense heat. It was heated to millions of degrees and then rolled over and over until the achievement of the highest level of perfection. It seems like a confession of a glassblower, but this is exactly how all celestial bodies are created.

The millions of craters on the surface of our Moon are not the result of it being bombarded by comets or asteroids. I do not know who proposed the idea that all those craters are footprints of alien bombardment because it seems that all the asteroids in space are magnetized to our poor satellite.

Along with extreme heat, any thermo reaction is accompanied by the emission of gases. During the satellite's gradual cooling, heated gases make their way to the surface where, after reaching a less dense environment, they burst into space *(Pic. 36)*.

Gaseous bubbles that did not make their way to the surface were trapped beneath the already cooled down crust, creating myriads of cavities of various sizes. We can observe the same effect in cheese where all the cavities of sliced cheese have spherical shapes. There is no triangular or square bubble in it because, at the time of its creation, the cheese had a liquefied condition.

(Pic. 36)

As in the diagram above, new bubbles always move to the surface where, after reaching a less dense environment, they eventually burst. We observe this process in our everyday life when we cook thick porridge. As a side note, the depiction of the satellite next to the diagram above is nothing other than a picture of the surface of yogurt. Yes, regular yogurt we buy in grocery stores. The surface of yogurt is very similar to the surface of our moon, yet nobody ever bombarded the surface of our beloved dairy product. The problem with our perception of the world is that we cannot imagine that this process can flow in such enormous proportions.

From times of antiquity until the Apollo Lunar Missions, there were many accounts of vapour clouds emitting from the surface of the Moon, a perfect indication that our Moon is still cooling down. There are many so-called mysteries regarding our satellite. The Lunar Mission's statement: "The Moon rang like a bell" is absolutely right. Of course, we cannot expect it to have a perfectly spherical form inside, but the Moon and all the celestial bodies in the universe are hollow indeed. That is why comparing our satellite with a "bell" is very plausible. During its creation, the heat emitted was so immense that it basically moulded an enormous rustproof metal ball, and of course any substantial impact would produce a bell effect.

It reminds me of NASA's 2009 LCROSS mission to the Moon when it purposely impacted the surface of the Moon with one spacecraft while the following spacecraft was analyzing the plume that originated from the first impact. And the mission was designed to search for water on the Moon. While the entire planet was watching the completion of this so-called significant event, many people, including me, were puzzled by NASA's water-searching methods. I was surprised at how any earthling could take such responsibility to conduct such a dangerous task for our satellite.

Considering that our Moon, like any celestial body in the universe, has a hollow body, I would not risk conducting even fireworks there. Our lives entirely depend on our satellite and any substantial change on it will immediately be reflected on our planet. Below are two such unfortunate scenarios:

1.   If the moon flies away then after unpredictable perturbations, our planet will return to a no-tilt position, which will result in occurrences of Ice Ages.
2.   If it disintegrates and eventually bombards the earth with its debris, then there is no point in writing this book since there will be no survivors to ever read it.

Knowing that the enormous basin of today's Pacific Ocean is just a crater formed due to the ejection of a satellite, we can try to find the runner. It is a celestial body, which looks very similar to our moon, has the largest orbital eccentricity and has the closest to the sun orbit — Mercury.

If we go back to the illustration above (Pic. 35), the satellite on the left is planet Mercury, which is a satellite of Earth by birth and sibling to our Moon.

The throwing force of the Earth reversing was so immense that the satellite escaped from Earth's gravitational pull. And where else could it go except the orbit of our Sun? Of course, a substantial role in this was played by the direction of ejection and — apparently — it was directed towards our Sun. Compared to nearby planets, Mercury has quite an impressive size and escape velocity. Just the passing of such a fiery planet through orbital trajectories of other planets could cause immense destruction and orbital turbulence among them. But if we accept Immanuel Velikovsky's theory about Venus' historically recent appearance, then there was no other planet on the path of Mercury's gradually shrinking orbit.

After ejecting such a substantial mass into space, the Earth itself was pushed away from the Sun. According to Mercury's elliptical orbit we can only guess if the paths of parted "mother and son" have ever intersected again. Considering Mercury has the largest orbital eccentricity in the solar system, with every Mercurial year its eccentricity should lessen and lessen until it achieves a perfectly circular orbit. It would be quite solvable for our astronomers to calculate Mercury's point of separation from Earth, which would help us determine the time of its occurrence and the proximity of our planet to the Sun right before the cataclysm.

Let us get back to the time of satellite ejections again. Right after the "delivery," our planet lost its first offspring forever. I think it would be harder for Earth to overcome this tragedy if its satellite were not adopted by the Sun. After all, Mercury was adopted by its grandparent. But life must go on, and along with tragic loss there was a second "baby" to be born. The basin of the Indian Ocean can be asserted as the birthplace of Earth's second offspring — our Moon.

In 1864, when zoologist Philip Sclater wrote his article about the puzzling similarities among fossils and living mammals of Madagascar and the Indian peninsula, he came to the conclusion that these two lands were linked together by a now sunken continent, which he called Lemuria. And he was not very far from the truth but for one misconception: the so-called continent of Lemuria did not sink; it was ejected to a near-Earth orbit.

After announcing the Lemurian concept, many geophysicists erroneously decided that the Indian plateau had separated from the island of Madagascar and gradually shifted away towards the Himalayas. On my depiction of the Indian Ocean *(Pic. 2)*, you will notice that there is no peninsula of Hindustan. I purposely erased it from that period of time because the Indian plateau as it is now was formed by an enormous landslide from the Himalayas. The ejection of the moon pulled behind itself masses of earth causing massive landslides. It is not coincidental that the Indian plateau points towards the centre of the Indian Ocean. It points to the point of ejection.

Now, knowing the answers of satellite creation we can surely assert that: **Neither the Moon nor any other satellite can be captured by its parental planet. Satellites are products of the planets themselves.**

Many of us have seen pictures of the famous (completely out of place and style) "sphere within a sphere" golden complex located in the Vatican and other countries around the globe *(Pic. 37)*. What catches the eye here is that such a mysterious and modern looking complex is located in the middle of the entirely classically decorated architecture of Vatican City.

(Pic.37)

The birth of the satellite from the planet and apparent crack on its back perfectly matches the mysterious meaning of these spheres. It is either a strange coincidence or someone's silent announcement of this hidden knowledge.

Mars

As our closest and most mysterious neighbour, planet Mars must be discussed separately.

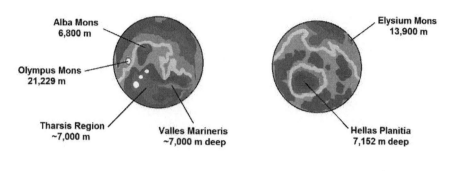

(Pic. 38)

A topographic map of Mars indicates that its entire northern hemisphere is lower than the rest of the planet by 1 to 3 kilometres. That distinctive feature is called the Martian Dichotomy.

Knowing the catastrophic events that took place on Earth, the reader of this book can easily detect very obvious feature similarities between our two planets. Here is the chart of the Martian equivalents of Earth's landscapes.

Mars	Earth
Northern Hemisphere	Pacific Ocean
Valles Marineris	Atlantic Ocean
Hellas Planitia	Indian Ocean
Elysium Mons	Iceland

As we can see here, the catastrophic events of satellite ejection on Mars have flowed by the same scenario as on Earth.

- In order for a satellite to obtain the velocity necessary for ejection, the surface of the part ready to separate from the planet has to be more than half the size of the entire surface of the planet. We can see it on both planets. And the Martian northern hemisphere and Earth's Pacific plus Atlantic Oceans cover more than half of the entire surface of the globe.

- In both cases the separation led to the formation of a planetary scale crack. On Earth, it led to a 5,000 kilometre displacement of crust (the Atlantic Ocean). Mars, on the other hand, managed to get just a 200 kilometre-wide crack (Valles Marineris). And that played a fatal role in its destiny.
- The ejection of the second major satellite from Earth left behind it the crater of the Indian Ocean, whereas the Martian crater Hellas Planitia is also the result of an ejected second satellite.
- Compared to Earth's Iceland, Mons of Olympus, Alba and the entire Tharsis region of Mars have much larger scales. This fact indicates a completely different fate of the ejected satellite.

Considering the absence of decent size satellites around Mars, and such a large quantity of shattered material in the neighbouring asteroid belt and on Mars itself (Tharsis region), we can make the assertion that the birth of the first satellite was not successful and both of them — mother and child — died. And again, all the volcanoes on Mars are the debris from the satellite ejected to space with the only difference from Earth being that the Martian volcanoes (the largest in our solar system) are just fallen back shatters of the first satellite, and all the members of the asteroid belt may be debris of shattered Martian crust.

Comparing the surface areas of the only perfectly spherical planetoid in the asteroid belt, Ceres (2.850.000 km²), and the Martian crater of Hellas Planitia (approx. 16.610.600 km²), we cannot eliminate the possibility that Ceres could be the second orphan satellite of Mars. The devastated planet had lost its power to hold its second offspring (Here we should not eliminate the possibility of accidental collision of two siblings during their take off).

As we can see, our two planets have very similar fates, except in the case of Mars something went terribly wrong. The discrepancy of mass and inertia of the first satellite brought to failure in the formation of a sphere, which led to its disintegration and massive hailing back to Mars. Exactly the same process occurs when we try to make a soap bubble. A successful bubble has a very smooth and gracious movement, but when it bursts, all the former lightness momentarily disappears and its debris rashly smashes back to the ground. That is exactly why Martian volcanoes are the largest in our solar system. Basically, due to the unsuccessful formation of the first satellite, the stratum torn away from the entire northern hemisphere smashed back onto Mars again.

If during the births of Earth's satellites there were very few survivors, Martians on the other hand did not have any chance to survive. Because of unimaginably high temperatures, the Martian atmosphere instantly evaporated. A much smaller comparison can be made with the disaster of a space craft during liftoff. As we had such accidents in history, we know they destroyed the launching platforms.

The Martian surface shows distinctive differences in the quantity of craters of the two hemispheres. Of course, all the debris bombarded the Martian surface quite evenly. But there are two different effects between a fiery ball smashing into the ground, and splashing into molten lava or water. The heat emitted during satellite ejection would reach millions of degrees and any bombardment of liquid substance in the form of drops of the same substance would momentarily dissolve and blend with the rest of the landscape. The rest of the camouflaging process would be accomplished by waves of rapidly evaporating water.

Along with the similarities in fates between our planets, the Martian landscape does not show the remaining causers of separation — mountain ranges. Unlike Earth, with its north to south mountains on the American continent, Mars does not have any continuous mountain ranges. According to our present knowledge, we can point out two possible answers:

1. Either Mars never had any mountainous causer of separation (very doubtful)
2. Or the Mountain range was taken with the ejecting satellite and that overweight played a fatal role holding the satellite back from forming a sphere

The absence of any other cause for the satellite ejection (formation of its own "Vampire Ocean") makes us lean towards the second variant.

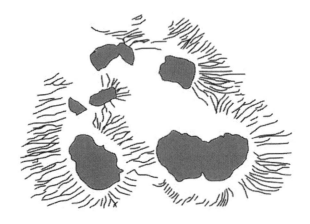

(Pic. 39)

High resolution images of Mars prove the existence of water in the past. The washed away crater silhouettes on the bottom of Valles Marineris *(Pic. 39)* indicate that a substantial quantity of water flowed into the crack after or during the bombardment of falling debris of the satellite. Lines directed to the centre indicate the direction of water drainage into the cavities of the fractured crust. Exactly the same effect can be observed in our everyday life, where, despite the spinning character of draining water, the lines of the sediments are always directed towards the centre of drainage. All the hills in the centres are water passages overfilled by debris brought by receding waters. It is the same pattern we see in the over spilled drainage of our large cities during floods.

According to images taken on the bottom of Valles Marineris (one of the deepest points on the Martian surface) we can assert that along with the molten lava of the Tharsis region, the fracture of Valles Marineris was also filled with water, which means that not all Martian waters have evaporated into space. Considering that all celestial bodies are hollow, the water could leak into the interior cavities of the planet.

It does not mean that water had leaked all the way to the centre of the globe, but it could easily fill up the interior walls of the Martian globe. The only possibility for that to happen would be the planet's complete stop of rotation. Knowing the exact cause of the Martian catastrophe we can surely assert that such a massive impact of the recently ejected satellite could easily knock the planet off its ordinary rotation and bring it to a complete stall. The absence of a centrifugal force would lead to the sudden shrinkage of the globe and the creation of an enormous suction force directed towards the planet's interior. The gradual acceleration of rotation could pump some water from the interior cavities of the planet towards its exterior, but considering the boiling conditions and very weak magnetic field of the post-cataclysmic Martian surface, any water would momentarily evaporate into space.

Comparing the cooling down ratios of Earth's largest volcanoes with the Martian Olympus Mons, we can try to create a chronology of events. Taking into consideration that all the largest volcanoes on Mars are now extinct, we can surely assert that the catastrophic events of ejection of the Martian satellites took place way before our own. At the same time, active volcanoes on Jupiter's moon "Io" indicate its comparatively recent birth.

Satellites

Now, knowing the processes which take place during the birth of satellites, I would categorize them as the following:

- Spherical satellites which were ejected from the planet tangentially (Mercury, for example)
- Spherical satellites ejected at the right angle from the surface of the planet (our Moon, Mimas, Tethys, Lapetus)
- Irregular shaped satellites shattered from the birth of the main satellite (satellites of Mars)

In the image below *(Pic. 40)* there are two of Saturn's satellites: Mimas with its huge crater and five-kilometre-high walls, and Lapetus with its thirteen-kilometre-high equatorial ridge.

Saturn's satellites – Mimas and Lapetus

(Pic. 40)

The main crater on Mimas is the most clearly visible feature even though every satellite has the same typical feature. These huge craters are just traces of enormous bubble-bursts frozen in time. But those "birthmarks" of every satellite can have different origins. As we know, the main role in the process of the moulding of any satellite is played by the wave. As we can see on Mimas, a wave travelled across the entire globe and cooled down right before collapsing in one spot. On Lapetus, on the other hand, a wave originated in one spot, began to travel away and cooled down somewhere in the middle of the globe.

Such a large crater as on Mimas can be seen on almost every satellite. This unique feature can be considered as the "belly button" of satellites. During its mould, the upcoming scars left from the bursting of molten bubbles overlap the previous ones, eventually camouflaging the satellite's "belly button."

Knowing the processes of satellite creation, it is much easier to understand our solar system. Everything is quite obvious; planets with high rotational speeds have the most satellites. Planets with low rotational speeds have the least satellites or none. That is why such "sprinters" as Jupiter and Saturn have the most satellites where most of them were ejected tangentially.

In modern astronomy, there is a division between inner and outer planets of our solar system. Inner planets are considered as terrestrial (composed of rock), and outer planets like Jupiter and Saturn are composed of gas. Knowing that all the satellites come from their mother planets, such terrestrial satellites of Jupiter as Io, Europa, Callisto and Ganymede could not be created from gas only. Comparing the main satellites of Jupiter with the satellites of Earth, we can clearly see an indication of different birth conditions. At the same time, comparing the landscapes of Mercury and the Moon with satellites of Saturn we can see a clear similarity between them. There's a clear indication that planet Saturn had gone through the same process of satellite creation as planet Earth.

The other and very important feature of satellites is the direction of their rotation around their mother planet. Along with satellites rotating around their mother planet in one direction, there are satellites rotating in the opposite direction also. For example, Jupiter, Saturn and Uranus have satellites rotating in both directions. So, how can a planet rotate in one direction and eject satellites in the opposite direction? The impossibility of such a task brings us to the conclusion that **the law of planetary reversals is universal** and all the planets do experience recurrent reversals. Identical to our Earth, all the planets in the universe undergo the same processes of creation, which means the formation of their own "Vampire Oceans," the eventual tilts of the planets and ejection of their own satellites. Laws of physics are the same throughout the universe.

Planets and Stars

In the chapters above I tried to prove the appearance of satellites and the impossibility of them being adopted by neighbouring planets. But what about the planets themselves?

In his book *Worlds in Collision*, Immanuel Velikovsky asserts that many calamities of our civilization in the middle of the second millennium BC were due to the appearance of the "Morning Star" — Venus. I would call it the "newly baked Venus." For us humans with quite a short lifespan and memory, witnessing the birth of Venus gives us a very important clue in understanding the main principles of our universe.

The present temperature on the surface of Venus is 462 °C. The reason why I mention the word "present" is because these numbers will gradually subside. Venus, at present, rotates around its axis in 243.02 Earth days and in the opposite direction from the majority of planets. Why at present? Because this number will also gradually change. Comparing the rotation periods of Earth and Venus we could say that Venus is almost motionless. And it is quite explainable. In contrast to a boiled egg, a raw egg does not spin. The entire planet is still too liquid to speed up its rotational period. The first question that comes to mind is: What about our planet Earth, with its borderless and liquid oceans? If we compare two barrels filled with water, one of them made of metal and the other from rubber, the metal pail would be quite easy to roll. Flattened down by the weight of water, the rubber barrel would be impossible to move. In contrast to liquefied Venus, planet Earth has a solid shell.

The opinion that planets are just sparks from the fiery Sun is absolutely right. I remember that as a child I was collecting tiny metal balls left from electrical welding. Most of them had perfectly spherical shapes, and what was more intriguing was that most of them were hollow. As we know, the main principle of arc welding is the difference of electrical poles where a short circuit creates a spark.

If there was an immense spark on the Sun, which led to the appearance of Venus, then there was a cause. A suddenly discharged spark could not be caused by nothing. No comet or asteroid passing by would be able to provoke such a reaction of the Sun. In order for a welder to get a good spark, the ground line should be properly secured to the object. In other words, there should be substantial resistance of opposite polarity.

It would be wrong to consider our Sun as just a huge sphere of flame. It must have its polarity also. In his works, Zecharia Sitchin asserts that our solar system has one more planet in its family, which visits us every 3600 years. If we are to suggest that our solar system has a binary structure, where two stars have highly ellipsoidal trajectories, then every close passage of the stars around each other should provoke an immense electrical discharge, and the birth of another planet or planets accordingly.

In the same manner, in order for our Sun to eject a Venus-sized spark, there should be a close passage of a celestial body that is substantial in size. The existence of a second star in our solar system is very probable and can be proven without even searching for it in the sky. The existence of so many planets in our solar system indicates the existence of an external spark. Much like male and female, stars can be categorized by similar criteria. Reproduction involves both parents. If the star is single, there is no exterior catalyst and no planets around it.

Knowing the process of creation, it is easy to comprehend the basics of our solar system. All the planets in our solar system (except Mercury) are direct descendants of our Sun. After each passage of our second star, sparks appeared between two oppositely charged stars to create planets. Every passage of the stars around each other flows at a different distance. A closer passage means a more powerful spark and the appearance of a larger planet. A larger passing distance creates lesser sparks and the appearance of planets smaller in size.

But let us forget for a moment about the astronomical properties of our planets and try to look at our solar system as some kind of clock, where the closest planet to the Sun is our near past, and the farthest planet from the Sun is our distant past. If, after delivering Mercury and the Moon, our planet moved away to the next orbital trajectory from the Sun, it would be rational to suggest that with the appearance of every new member in our family, along with losing some physical part of it, our Sun loses its power also. Every short circuit of battery weakens its power.

In other words, with the appearance of every new member of our solar system, all the planets simultaneously shift away from the Sun, forming some kind of conveyer. Like in any family, the birth of a new member shifts the siblings to the second stage and the newborn member always gets the most attention. In this case, we can assert that Saturn is older than Jupiter, which is older than Mars. Planet Earth (along with Mercury and the Moon) is younger than Mars but older than the "newly baked" Venus. And the temporarily rejected planet (by our scientists) from our planetary family, Pluto, is one of the oldest members of our solar system, and considering the existence of satellites accompanying it, planet Pluto has the same origin as all the planets of our solar system.

Names	Equatorial Diameter (in relation to Earth)	Rotation period (Earth days)	Confirmed Moons
Mercury	0.382	58.64	0
Venus	0.949	-243.02	0
Earth	1.00	1.00	1
Mars	0.532	1.03	2
Ceres	0.07	0.37	0
Jupiter	11.209	0.41	67
Saturn	9.449	0.43	62
Uranus	4.007	-0.72	27
Neptune	3.883	0.67	14
Pluto	0.18	6.387	5

(Pic. 41)

If as proposed by Z. Sitchin, planet (star) Nibiru orbits the Sun every 3600 earth years, where every such passage is accompanied by the ejection of only one planet, then we could try to calculate the approximate age of our solar system. But knowing very little about the complete list of our planetary "siblings" it would be hard to accomplish the task. We have very limited information about planets located behind the orbit of Pluto.

But this is not the only criteria for creation. Let us look at the chart of planetary attributes *(Pic. 41)*. Here are just three of the most important ones: diameter, rotation period and quantity of satellites. Of course, it would be very helpful here to have the masses of the planets also, but unfortunately, we are not capable of calculating real masses of any celestial bodies (yet). Present calculations of planetary mass are wrong and have been determined by the volume of the spheres. If somebody could tell us the volume of cavities inside of each planet, then there would be no problem calculating (at least) its approximate mass.

- If all the planets of our solar system are the offspring of our Sun, then planet Mercury should not be considered a planet at all. According to its satellite-type origin and present orbit

110

around the Sun, it is adopted by the Sun as a "satellite of Earth" and — as with all the satellites — it is infertile.

- Newly-baked Venus, the youngest member of our solar system, is in too liquid a condition to spin. Because of its external 462 °C temperature, it cannot obtain any substantial rotational period and cannot have any satellites (it is still too immature to have children)

- Earth — the second youngest planet — had delivered two satellites where one of them (Mercury) was adopted by the Sun

- Mars – the third youngest planet — had delivered two satellites with one of them unsuccessful and the second one (presumably Ceres) gone to the nearest orphanage (asteroid belt)

- Starting from Jupiter we can see that larger planets having higher rotational speeds have more satellites. Smaller planets spinning at a much slower rate have lesser satellites accordingly.

Here it may be reasonable to assert that during the creation of Saturn and Jupiter, the distance between the two stars orbiting each other was minimal and led to a more powerful electrical discharge. However, the problem is that we do not know the actual sizes of gaseous planets.

Knowing the process of creation of Mercury, Venus, Earth and Mars, the outer gaseous planets are still a complete mystery to us. One thing is obvious: all the planets of our solar system are siblings (except Mercury). Children of the same parents are supposed to have the same genes aren't they? If all the planets of our solar system are creations of our Sun, then all of them should be composed of the same matter. That means that starting from the planet Jupiter, all the planets obtain unimaginably high speeds of rotation, lifting huge masses of dust and creating enormously thick layers of gaseous atmosphere around planets. According to that we can certainly assert that the real sizes of gaseous planets in fact are unknown. Let us just try to imagine a planet eleven times bigger than Earth, which makes one complete revolution around its axis of rotation in just 9.84 Earth hours. We should not forget here that these numbers came from calculations based on exterior dimensions of the planet, which is a layer of dust.

Have you ever seen the mixing of thick dough in an electrical mixer? With the mixer's high RPMs, the dough inside supposedly must spin with the same speed as the mixer itself. But in fact, the dough has much less speed on the outskirts of the pan. In other words, the real speeds of gaseous planets are much higher than we can observe with telescopes.

By considering our solar system as a clock we can make some major assertions. When Earth lost its first satellite (Mercury), it was adopted by our Sun. When Mars experienced a major cataclysm with the "miscarriage" of its first satellite, some of its debris and Martian second satellite (Ceres) — which is much smaller than Mercury — went in the opposite direction from the Sun and formed an asteroid belt. Of course, the first thing that comes to mind is the huge gravitational pull of Jupiter. But if we are to suggest that at the time of the calamities of Mars, Jupiter could have been located on the opposite side of the Sun, then it would have no influence whatsoever at the asteroid belt. Then why did all the debris of a once thriving planet form an asteroid belt instead of being pulled towards the Sun?

Of course, the main role in the future destination of satellites is played by the direction of their ejection. In order for Mercury to escape Earth's gravitational pull, it had to be ejected towards the Sun. In the case of Ceres, in order for it to travel towards the asteroid belt, the vector of its ejection had to be directed away from the Sun. But besides those conditions there should be something else.

The locations of the gaseous giant planets in our solar system indicate some kind of influential border of our Sun. Starting from the asteroid belt until the premises of Jupiter, the planets begin to obtain unimaginably high speeds of rotation. This effect can be compared with the gravitational pull of our own planet. During ascension to space, all the spacecrafts are in the gravitational pull of Earth until they reach the point of zero gravity. Of course, zero gravity does not mean absolute independence from Earth, and all manmade satellites continue to spin around the globe in near orbit space.

A blade of a boat spinning in water will accelerate its rotation as soon as you lift the engine from the water. If we move an already spinning object from a denser environment to a less dense environment, the inertia of the spinning object will accelerate the object's speed of rotation. Exactly the same effect can be observed behind the asteroid belt where planets accelerate their speeds of rotation because of zero gravity of the Sun.

Let us go over our outer planets one more time. As we can see, planets located farther away from the Sun have lesser satellites. Let us start from the absolute champion — Jupiter — 67, 62, 27, 14 and 5. Once again, if we perceive our solar system as a timing mechanism then we can certainly assert that the ratio of planets and their satellites was not always the same.

- First of all, as soon as the planet crosses the zero gravity line, an immense speed of rotation begins to inflate the planet's atmosphere like a "Switch Pitch" toy. If we compared the sizes of present beyond-zero-gravity Jupiter with Jupiter within zero gravity, then there is a big chance we would see it in the size of our present Earth.
- Secondly, after passing the gravity border, planets obtain a large quantity of satellites. After a while when planets eventually begin to slow down, they lose their power and satellites accordingly. At this point, the satellites go to the second "orphanage," located way beyond the orbits of Neptune and Pluto — the Kuiper Belt — a much more massive version of the asteroid belt we know. The quantity of asteroid belts can indicate the quantity of gravitational layers of our Sun.
- Along with slowing down and losing their satellites, planets begin to shrink in size.

A perfect explanation for the structure of our solar system is the rotational speed of its every member.

- The rotational period of Venus is close to nothing (a raw egg does not spin) and, after gradual cooling, planet Venus will eventually start to accelerate its rotation.
- Earth and Mars have substantial speeds of rotation.

- Farther from the Sun (which means older, accordingly) come the gaseous planets Jupiter and Saturn with their unimaginably high speeds of rotation.
- Following them is Uranus and Neptune, which have slower rotational speeds because the energy of rotation eventually decreases even in weightless environment.
- Located beyond Neptune, Pluto has the slowest rotational speed.

If, along with the birth of every new member of our solar family, all the planets shift their orbits away from the Sun, then larger-scale sparks create galaxies. And if we go farther, the entire universe was created respectively. The only question that remains is where is the second causer of the spark?

Let us imagine our travel to the remotest part of Earth to meet with the largest and oldest family in the village. Seeing an entire family gathered together always tempts us to find the oldest one in the clan. We never try to determine family members by their stature.

Our observation of the universe on the other hand has strangely different principles. Instead of determining the oldest constellations, we categorize galaxies and stars by their sizes. As on any major feasts, the patriarch of the clan always sits at the head of the table. His children sit next to him, followed by grandchildren and so on. Our universe has the same patriarchal structure; the centre of our universe is the forefather of all the galaxies, solar systems, planets and their moons.

According to modern science, the death of a star must be accompanied by its extreme expansion. If all the planets are sparks of the stars and basically composed of the same chemical elements as their star, then during the process of cooling, all the planets would supposedly shrink in size and not expand like balloons. The major misconception of modern science is the consideration of every celestial body in the universe as a solid piece of rock completely independent from laws of physics. We do not need to be scientists to assert that a figure skater spinning on ice will accelerate his or her speed of rotation as soon as he or she folds their arms towards their body.

For some reason, our astronomers disregard the basics of centrifugal force in space. Therefore, we have to admit here that, unfortunately, all our scientific data regarding the masses of celestial bodies has to be reconsidered, and our astronomical calculations of planetary mass based on radius and weight of cubic metre of rock are erroneous.

If, during their formation, thanks to the force of surface tension, all the satellites were hollow, it would be quite logical to assume that all the planets are hollow as well. Of course, the thickness of the shells of satellites and planets are incomparable.

Knowing about the birth of planets, we cannot miss another important aspect. Satellites are getting ejected from planets due to their recurrent reversals. Planets on the other hand are appearing from the sparks created between two celestial bodies passing too close to one another. Despite different causes of birth, all celestial bodies obey two universal laws: surface tension and centrifugal force. Thanks to the surface tension effect, all celestial bodies have spherical shapes. Gases emitted from thermo-reaction inflate celestial bodies from inside.

As the saying goes, there is no smoke without fire. Just the fact that people came to the idea of a hollow Earth means that our ancestors knew something very important which we do not. What if our planet has an enormous cavity inside? In that case, we cannot eliminate the possibility of the existence of a star still burning in the centre of the globe.

The only question remaining is how to get there. In the face of an approaching imminent catastrophe and presumably having the technologies to do so, some ancient civilizations could have been able to escape an upcoming catastrophic event of ejection, where the safest place to be would be the interior of Earth. If some of our ancestors made their trip towards the inner surface of Earth, then independent of their location (whether it be on Earth's outer surface or on its inner surface), all the living things on Earth should have a fairly similar appearance with the only difference being that of skin colour due to exposure to the lights of different suns.

The next questions to arise are: First, how long would it take for Earth's interior Sun to burn all its resources? And secondly, can there be any water?

- As we see, the cooling down of active volcanoes is quite a long and continuous process, and the immense heat of molten rock does not cool down so fast. Thermo reactions flowing inside somehow sustain a steady temperature for millennia.

  According to modern science, our moon is 4.5 billion years old. Considering our civilization's verbal memory, the age of our Moon cannot be older than 20,000 years. I am not a good mathematician, but if the Moon's 4.5 billion years is equal to no more than 20,000 years, then how old is our Earth?

  If, after thousands of years, we still have erupting volcanoes, then it would be logical to assume that a satellite-size star can burn even longer. It can point to the high possibility that our Earth still has an active interior sun.

- According to modern science, water is a unique compound, inherent to our planet only. But if planets are sparks of their stars, then it would be logical to conclude that the composition of planet Earth is equal to the composition of our Sun, which basically means that the entire Mendeleev's Periodic Table of planet Earth also exists on our Sun. Even water. If there are molecules of hydrogen and oxygen on the sun, then why can't they form water after the cooling?

Knowing the cause and process of creation of planets and satellites, we can certainly assert that every spinning celestial body, including our Sun and all the stars in the heavens have hollow structures. I found a few very curious videos on the Internet depicting our Sun being hit by an asteroid. The most intriguing part was not the impact itself. After smoothly entering the Sun's body, it caused an enormous blast on its way out, on the Sun's opposite side. The first impression we get is that it is a hoax video, but in fact it proves the hollowness of the Sun.

Experiments with a water bubble in space show that by inflating air into a drop of water, it can remain trapped inside forever. In contrast to a water bubble, our Sun is one enormous sphere of fire with constantly flowing thermo reactions in it. Even if we are to suggest that at the time of its birth our star was one solid piece of molten lava, then after an x-amount of time, due to centrifugal force and constant emission of gases, it gets inflated regardless. All the gas bubbles located close to its external surface are bursting towards the outside. Gases generated close to the interior walls will find the shortest way out from the boiling mixture and therefore will head towards the less dense interior of the star.

Considering that after thousands of years from the ejection of Earth's satellites, volcanoes of our planet are still active, we can try to calculate the age of our Sun. The parallel I am trying to draw here is that the birth of our Sun is no different from the birth of our Moon or any other satellite. The only difference is in their sizes. If it took (let us suppose) 15,000 years for our Moon to cool down, then the cooling ratio of our Sun should depend on the size difference between our Sun and our Moon.

Like everything else in the universe, galaxies and solar systems eventually burn all their fuel and die. But the death of a star does not mean its rotational stop. It is the other way around. Much like a figure skater spinning with tightly folded arms achieves maximum speed of rotation, the shrinkage of stars accelerates their speed of rotation. Here we must mention that the acceleration process of a planet with a cooled down hard crust leads to the expansion of its atmosphere with lifted up dust and lighter particles of soil (examples are the atmospheres of Jupiter and Saturn).

The shrinkage of a gradually diminishing thermonuclear reaction of a star will lead to its rapid acceleration of rotation. So-called Black Holes in space are nothing else than gradually extinguishing stars and galaxies spinning at a constantly accelerating speed, sucking all matter into them like an enormous whirlpool. And who knows what happens when all the matter collides in one point. It may burst in every direction again, creating new galaxies and stars, and that process of birth and death can go on infinitely.

After consideration of the main principles of the universe, the Biblical saying "From dust we came, to dust we will return" can be relevant not only to humans and all the living things on Earth. It can be universal, literally universal.

By exploring our own solar system, we can certainly answer the question of the creation of our infinitely expanding universe. Like everything else, our universe lives by a conveyer-type principle. The disintegration of a much bigger star was the birthday of our Sun, which will eventually adopt the same fate.

Every newborn planet thrives in his "parental house" until it moves on its own (beyond Sun's zero gravity orbit), achieving its maximum velocity and having its own children (satellites). Time goes on and a planet reaches its forties, where senility begins its slow and irreversible course. When planets begin to lose their velocity and power, satellites begin to leave their parental nest to continue their own voyage into the suburbs of the constantly expanding solar system. In the end, planets shrink in size and lose their velocity to barely sustain their own living. After the complete stop of their hearts (stop of rotation), they just calmly flow in a suburban graveyard of once mighty planets.

As everything else, every living substance in the universe has its peak and fall. The flourishing of giants in the past, a much longer lifespan of our ancestors and such a large variety of diseases among all living things at present, indicate that we live in an environment fundamentally different from the original. Unfortunately, we must add here that we live in the declining stage of our planet. Our Earth has two adult children (the ejected satellites are comparably cooled by now). In comparison to our lives, our planet is definitely past its forties.

Once again, I would like to get back to Karl Brugger's book *The Chronicle of Akakor,* which tells the history of the Ugha Mongulala, an Indian tribe living on the highlands of the Andes from the time of arrival of the Former Masters (gods), which, according to their calendar, took place around 13,000 BC. The high priests of Akakor describe two great catastrophes in the history of their people that happen approximately 6000 years apart. According to them, the first great catastrophe took place in 10,468 BC.

The first Great Catastrophe gave the surface of the earth a different shape. The course of the rivers was altered, and the height of the mountains and the strength of the sun changed. Continents were flooded. The waters of the Great Lake flowed back into the oceans. The Great River was rent by a new mountain range, and now it flowed swiftly toward the east. Enormous forests grew on its banks. A humid heat spread over the easterly regions of the empire. .....

(Page 39)

There is no doubt that the "flowing of waters of the Great Lake (the Vampire Ocean)" into the oceans and the "change in the strength of the sun" proves that the period of time described in this citation is of the ejection of satellites and the following flood.

Here are even more intriguing citations from the same book (Page 140) with my analyses of each event:

- *One of the maps shows that our Moon is not the first and not the only one in the history of the earth.*
  ➢ They are absolutely right. After ejection of Mercury, the Moon is the second satellite.

- *The Moon that we know began to approach the Earth and to circle around it thousands of years ago.*
  ➢ Definitely. Ejected into space, the satellite could not stay at its maximum distance from the Earth and gradually it is supposed to adjust its gradually diminishing orbit.

- *At that time, the world still bore another face. In the west, where the charts of White Barbarians only show water, was a large island. And a gigantic mass of land was in the northern part of the ocean as well. According to our priests, these two were buried under an enormous tidal wave during the first Great Catastrophe, the war between the two divine races.*
  ➢ Of course, the pre-ejection period of the present area of the Pacific Ocean could not be covered by water only, and of course there could have been a large island and

"gigantic mass of land in its northern part" because the existence of such a massive barrier on the way of a rapidly traversing continent is a crucial condition for it to obtain the necessary momentum for ejection. These two (*large island and gigantic mass of land*) were swept and rolled away by an enormous tidal wave. Entire strata of the former "Vampire Ocean" along with its thriving flora and fauna were transformed into an enormous fiery ball, which we (White Barbarians) call planet Mercury. During the reversal preceding the appearance of the "Vampire Ocean," our planet turned into a position where half of the globe began to lose its vapour and the other half was gaining more and more water. Such an uneven distribution of water around the globe could not flow painlessly. If our civilization is constantly fighting for fuel deposits, imagine our planet divided into lakes in one hemisphere and deserts in the other. Outbreaks of wars would be inevitable. And, of course, compared to the waterless lowlands of the globe, the evergreen coasts of the expanding "Vampire Ocean" would be considered as a climatic paradise. Envy on the part of starving lowlanders would inevitably bring the world to war. (Here we should not eliminate the possibility of the metaphorical meaning of the two divine races, which are water and fire).

- *And they add that this war did not only lay waste to the Earth, but also to the worlds of Mars and Venus, as they are called by the White Barbarians.*

  ➢ The existence of Venus at the time of Earth's calamities is not certain, but the "war" between fire and water indeed had lay waste to Mars.

In the chronological table of events put together by Karl Brugger, gods left the earth in 10,481 BC, and the Great Catastrophe took place in 10,468 BC. Life on such a "time bomb" as Earth in that period would be too dangerous for the gods, and they departed from Earth years before its catastrophic reversal. It would take years for Earth to relatively calm down in order for the gods to decide to return.

According to the Ugha Mongulala, the second Great Catastrophe —
which took place about 6,000 years later — was connected with the falling
of a star.

> - *And the gods were grieved. Their hearts were filled with*
> *sorrow for the wickedness of man. And they said: "We will punish*
> *the people. We will eradicate them from the earth — man and*
> *cattle, the worms, and the birds in the sky — because they have*
> *rejected our bequest." And the Gods began to destroy the people.*
> *<u>They sent a powerful star whose red trail covered the whole sky.</u>*
> *And they sent fire brighter than a thousand suns. The great*
> *judgement began. For thirteen moons, the rains fell. The waters*
> *of the oceans rose. The rivers flowed backward. The Great River*
> *changed into an enormous lake. And the people were destroyed.*
> *They drowned in the terrible flood.*

> *Karl Brugger,*

The Chronicle of Akakor (Page 46)

Despite similarities between the stories of the Ugha Mongulala and
the Old Testament, I would like to emphasize the <u>powerful star with the
red trail</u>. Whether it be Z. Sitchin's planet Nibiru or I. Velikovsky's *Birth of
Venus,* is not important. A "powerful star with the red trail" could have
been a description of Nibiru and Venus equally. The main importance here
is that all three authors describe the same period of time in the history of
our planet. If the arrival of our second star provoked such a sparking
reaction of our Sun, then thirteen moons of rains on Earth sounds quite
plausible. The existence of so many witnessed chronicles describing the
same occasion proves its occurrence. If we could determine the real age of
our solar system, then it would be quite possible to predict the birth of the
next sibling in our solar family, where every newborn planet pushes the
elder siblings away from the Sun.

During the birth of Venus, the spontaneous shift of the planets has
changed their climates accordingly. On Earth, two major players of
destruction would be the oceans and the Moon.

- The sudden shift from the sun would create enormous tsunamis.
- The distance between the Earth and Moon would either suddenly increase or decrease, depending on the Moon's location during the shift. In both cases the Moon would stabilize its distance from the Earth after many fluctuations, where each approach to the Earth would generate enormous waves.

Now, by rediscovering our Earth's recent past and analyzing human history we can certainly assert that the Ugha Mongulala and other ancient tribes living on high altitudes of our planet are the oldest civilizations on the surface of Earth. And, as every generation teaches its descendants to respect its elders, the Ugha Mongulala deserve to be respected at least for the age of their civilization.

Luckily, not everyone perished in those catastrophes. In contrast to the people of the Ugha Mongulala, survivors scattered around the globe had degenerated from thirst and starvation, and had to restart their history from the beginning. The only ones who would have little chance to survive as a civilization are people living on the high altitudes of the Andes and Himalayas. Like Tibetan monks, Andean tribes cherish their silent history.

The entire book of *The Chronicle of Akakor* is dedicated to the history of ancient Indian tribes gradually losing territories to the constantly expanding postdiluvian civilization. Precious, partially revealed information in the book is not an accidental leak. Tatunca Nara (the leader of the Ugha Mongulala from 1968) seeing the desperate fate of his people decided to reveal some of the precious information to the world in order to save his tribe from complete extermination.

I do not know whether Tatunca Nara's tactic worked or not because humans are quite fast-spreading creatures. We constantly expand our habitats for two reasons: running away from others or looking for new sources of wealth. If people must destroy ancient ruins in order to build new structures, they usually do. They do this even though every clear-headed person realizes that the answers of our origin lay in the past and not in the future. The genome of survival constantly pushes us into the "bright" future and new territories.

Time relentlessly flows ahead, and with the birth of another member of our solar system the earth will be shifted to other orbits, and all the living things on it will experience the following shrinkage and mutation again. Our future midget generations will unearth our remains and be puzzled by our giant and alien origin.

By gradually developing our story, we can conclude that the story of the Great Flood described by all the cultures on Earth, could not take place at the time of the First Catastrophe (the ejection of satellites) and there are three basic reasons for that:

1. Considering the scale of such planetary destruction, there would not be many survivors to retell those events to their descendants. In a matter of minutes, 5/7ths of the planet, along with mountains, forests and every living thing on it were turned into ashes and ejected away. The miraculously-surviving population of the remaining 2/7th of the planet was destined to degenerate in order to survive.

2. It would be absolutely impossible for Noah's Ark to land on a "150 day old" active volcano. If it did not burn miles away from the volcano, then poisonous vapours would definitely kill everybody on board.

3. According to Z. Sitchin, right before the deluge, gods ascended to a near-Earth orbit and were watching the terrible events from their spacecrafts. Being on a near-Earth orbit during the First Catastrophe could mean suicide.

Many ancient myths and legends tell about a great worldwide flood and nothing about the suddenly appeared satellites. Surely during the First Catastrophe, the Earth experienced a worldwide flood, but survivors of the flood could not miss such a global event as the destructive ejection of satellites. But who could survive such an event? We can answer that question by eliminating the areas of most destruction.

- North America was too close to the pole and was not suitable for living before the cataclysm.
- South America was torn apart from the African continent in a matter of minutes, but according to the extraordinary story of Tatunca Nara his tribe survived it in underground caves.
- Antarctica and the African and Australian continents were the first victims of the ever destructive waves of the former

"Vampire Ocean." The survival of any living creature on those two continents could be considered a miracle and short-term.

- Europe and the northern part of the present Eurasian continent were also too close to the previous pole. During the Catastrophe, inhabitants of the central regions of Eurasia were buried under the thick layer of mud and icy waters of the enormous wave from the north.
- Tibetan and Himalayan caves could be the safest places to survive.

From the criteria mentioned above, we can assert that the cradle of the present civilization is located in the high altitude regions of the Andes and the Himalayas. The Meso-American civilization had spread its roots on the American continent and the Himalayan civilization began to spread in Eurasia.

One of the best silent snapshots of the Meso-American people who used to flourish in the pre-Columbian epoch is the Olmec heads, plentifully scattered on the Central American isthmus. Artistically crafted on spherical stones (remnants of terrific events of satellite ejection), Olmec heads bear the features of a Mongoloid race. If far ancestors of the Indian race belonged to the Mongoloid race, then the first question that comes to mind is: Where did they go?

The explanation is not in their disappearance but rather in their gradual mutation. During the time, dislocated from their original continents (along with every living thing on it) they began their gradual adaptation to the new conditions. Meanwhile, survivors of the more stable Eurasian continent have not changed so much.

One of the little known archaeological sites of the antediluvian epoch is the so-called Hal Saflieni Hypogeum, located on the Island of Malta. Archaeological excavations of this prehistoric underground site were accompanied with the discovery of nearly 7,000 human remains. Apparently, in its last minutes, the presently underground structure served as a refuge from the raging disaster. Unfortunately, it became a graveyard for people who were trying to find refuge in it. There are four very curious details about this site:

- All the chambers were filled within a short distance to the roofs with a mass of reddish soil.
- All the skulls of the excavated remains had an elongated shape.

- Some of the best-preserved skulls, which were exhibited in the Archaeological Museum of Valetta until 1985, have disappeared without a trace.

Let us scrutinize these details one by one:

1. The present location of that underground site and the fact that the entire structure was filled with soil indicates the occurrence of a flood. While trying to find refuge, people were caught by the approaching water. Here we must clarify that this archaeological site cannot belong to the period prior to the ejection of satellites. No living creature would have any chance to run to a nearby refuge. No building would be able to survive such an instant displacement of Earth's crust, especially located right in the middle of the Mediterranean Sea, an epicentre of crust displacement. The period of that tragedy most probably belongs to the Second Cataclysm on Earth, which is the biblical Great Flood.

2. An elongated shape of the skulls indicates that people were living in an environment different from now. Findings of similarly elongated skulls in Peru are not coincidental because they belong to the race that survived the greatest cataclysm on Earth — the ejection of satellites. According to that we can make an assertion that during the period between the satellite ejection and the Great Flood, planet Earth could have been populated by people with elongated skulls.

3. The sudden disappearance of those artifacts proves their exceptional importance. Strangely, some people do not want those skulls to be linked with modern humans.

If we are to assume that the biblical Noah and the victims of the Great Deluge at the "Hal Saflieni Hypogeum" site were living in the same historical epoch and even the same year, then it would be logical to assert that the legendary Noah and all the people of that epoch could have the same, elongated type of skulls.

If biblical Noah and his family were the only ones who survived the Great Deluge, then the elongated skulls of the rulers of ancient Egypt indicate that they are descendants of Noah or other tribes that survived. The fact that the majority of mummies belong to high-ranking people of ancient Egypt led us to believe that the Egyptian civilization was ruled by a race of aliens. But in fact, elongated skulls were typical for entire populations of ancient Egypt. The process of mutation does not take place in a matter of years; it is a matter of many generations. Every following generation was mutating in the mother's womb. And to expect an identical mutation of every newborn child would be erroneous. Every child has his own genes, timing and food ration during his development. That is why there is no consistency in skull types among mummified remains of the ancient pharaohs.

Researchers claim that the naturally elongated skulls that were discovered have twice as big a volume than modern people. If there was no necessity to have a much bigger brain then we would not have it. In other words, bearers of a larger brain should have had more abilities than bearers of a smaller brain. We are not talking about a primitive ratio between body weight and the weight of the brain itself (complete nonsense). We are comparing brain sizes and capabilities of their bearers — us and antediluvian people. One simple question arises: If modern people can do this much, then what were people with elongated skulls capable of?

As we know, some of the largest dinosaurs on earth had the smallest sized brains; something like a tennis ball. Despite that disadvantage, they flourished on our planet for many millennia. Today, in an era of constantly developing technologies, the shrinking of computer dimensions does not mean their lesser capabilities. They are much smarter, faster and smaller than their predecessors. Then why do we think that the bearer of a smaller brain has limited abilities? It's a paradox.

But there is another, hidden side of the coin. There are two different types of elongated skulls: natural and artificial. By binding a newborn child's head, it is possible to accomplish its gradual deformation. The main difference between them is the volume of the head. In the case of artificial elongation, the volume does not change. I believe, back then it would have been prestigious to have longer skulls because that distinctive feature would indicate a relation to a much older and respected tribe, and that was the main reason for the artificial elongation of children's skulls. That distinctive feature would open many doors for them into higher ranked positions in government.

Who are we?

Our perception of the environment surrounding us begins from the first days of our lives where our visual memory fixates on all the pictures and events we ever witness. That is why when we explore the universe beyond our planet, we always look for archived shapes and silhouettes familiar to our memory.

When we try to cultivate some vegetation away from its homeland, we always expect to see the same product, perfectly understanding that in order to get the same harvest, we need to provide for them exactly the same climatic conditions, altitude and water composition.

On any planet we explore, we expect to have conditions similar to Earth, with flora and fauna that are familiar to us. Let us take two similar and convenient for our habitat planets. Descendants of identical twins placed on two different planets will never belong to the same race again. When we find remains of giants, we automatically reject the possibility of our relation to them, but we are the same people with the same genes and same ancestors. The only difference is the environments we live in.

If we met Martians that had an appearance similar to ours, there would be a great possibility of our relation. In other words, a human race that can possibly live on other planets or even inside of our own planet can have different colours of skin, different stature, different voices, different everything. One thing is certain: every living thing in the universe is a reflection of its environment. The proof of that is the existence of races on the seemingly homogeneous surface of our planet.

The planets get through the same process of moulding as satellites and therefore interiors of planets also cannot be perfectly spherical. But even if we assume a perfectly round interior of the planets from their birth, we should not forget that the ejection of satellites fundamentally destroys all the evenness of the planetary crust. We have no clue how the thickness of the planet's crust can affect living things on its surface. A thicker crust means a stronger magnetic field. The suggestion that the conditions on the surface of a planet are identical regardless of the thickness of its crust would be erroneous. The thickness of Earth's crust in particular can play a substantial role in the mutation of all living things.

Mutation of living organisms mostly depends on structural changes of one of the main materials in the body — water. As mentioned above, all the planets of our solar system contain water, but every planet has its unique magnetic field, distance from the sun and rotational speed. Our expectations of finding water on different planets in our solar system with identical characteristics as on our own planet would be very unlikely. Molecules of water cannot have the same structure in different environments. Nothing can. Therefore, water brought to Earth from a different planet will automatically mutate to our planet's environmental conditions.

As many ancient sources say, people originated on planet Mars. I would not be too skeptical about that because, as we discussed earlier, planet Mars is much older than Earth. If we are to consider planet Mars as our past and planet Earth as our present, then planet Venus is our only future homeland. If humanity does not destroy itself before the cooling down of Venus, then we will have a chance to migrate to our next homeland. If we consider that Venus at present is located on Earth's before-satellite-ejection orbit, then after the migration of future "four feet tall" earthlings to Venus, the following generations of people will grow in size again. Our bodies will just adapt to the new environment.

Generations who will be born on future Venus will remember that their ancestors came from planet Earth and nothing more. It will be the same as descendants of people who immigrated to different countries heard from their parents about some distant homeland they came from. Some of them know some facts about the country of their ancestors' origin though many of them do not even know that country's whereabouts. Today, after hundreds of generations, human Martian origins seem like nothing more than myth, but myths cannot be created from nothing. Surely, if we do not know the origin of our appearance on Earth, it would be too much to expect us to know about the Martian origins of our ancestors. If someone never saw his childhood pictures, it would be difficult for him to recognize his own image when he gets older.

Gradually we conclude that the human race is a constantly migrating and mutating species. The genome of survival constantly drives us towards the exploration of new habitats. By migrating to different planets and galaxies we will constantly mutate and adapt to new environments without even noticing it. Again and again our future archaeologists will be puzzled by the discoveries of our considerably large remains and remnants of ancient cities. Much greater than our lifespan, recurrent cycles will always occur in our planet's future.

When we mention the word race, of course the first thing that comes to mind is colour of skin. We explain such a large diversity of skin colours by the exposure to diverse levels of ultraviolet rays, and common sense tells us that people living on the equator should have the darkest skin. But if we look at the dispersion of human skin colour around the globe we notice that the darkest people on earth are not concentrated on the equator at all.

If the African continent is populated by a black skinned race, then populations of equatorial regions of South America should have black skin also, but they do not. Considering that the South American continent was part of Africa, during their splitting, entire populations of both continents, especially African, were swept away. As we know, after the spontaneous shift of the entire American continent towards present west, the African continent was the first victim of two enormous waves smashed on to its shores (from Central America and Drake passages). In fact, it would be hardly possible for anyone to survive such an impact. If someone could escape the enormous waves from the west, then thermo-heat of the newly separated Moon would burn everything to ashes from the east.

Present aborigines of Australia also have one of the darkest skin colours on earth. However, considering Australia's low altitude terrain and very close proximity to the newly separated satellite (Moon), it can be considered as the second most dangerous place to be during the cataclysm. It would make perfect sense if the Australian continent was located on the equator then, but it is not. Indonesia is located on the equator and its population has a lighter skin distribution than Australia. Of course, we cannot eliminate the migration factor, but migration means movement from point "A" to point "B." Here it would be very important to determine the original location of point "A."

If we take into account the significant role of sunlight in the formation of skin colour, then the Eskimos of Siberia and North America should be the whitest people on Earth, but they are not. So, what if besides proximity to the equator and exposure to sun there are other factors affecting the colour of our skin?

As I mentioned above, before ejection of satellites the Earth was periodically experiencing Ice Ages. Extreme heat on the equator was forcing flora and fauna of the planet to migrate towards 45° latitudes of the northern and southern hemispheres, and that explains the appearance of diverse types of animals which strangely belong to the same family. If, for example, we visualize the situation when a flock of large cats was separated by one of these events, then following the isolation of formerly identical species for many centuries, could easily cause their transmutation to lions and tigers. And if we take into consideration that this process was repeated many, many times then we have quite a complicated collection of animals with their untraceable origins.

If, during the Ice Ages, the planet was divided into two corridors of life, then humans would have the same fate as the animals that were separated. And here again, two groups of humans isolated by different climatic conditions would develop completely autonomous racial features. The following polar shifts would join them together (of course not without fights between each other) in order to split them again during the next Ice Age. And our present civilization is a mixture of that complex process. Now, after knowing all this, I am wondering how the reader of this book will visualize Adam and Eve.

Considering that the habitats of the major races on Earth are located on different continents, there are also many racially mixed groups of people who live between the habitats of major races. The formation of these groups is influenced by:

- belonging to one of the major races
- relation degree to other major races
- habitat

By developing this topic I am attempting to introduce the idea that the appearance of sub-races is not just the result of interracial relationships but is mainly affected by their habitat on Earth. In other words, would it be after five, ten or twenty generations that people who migrated to different habitats eventually absorb the qualities of the inhabitants of that region. One of the main factors of race formation is the thickness of Earth's crust in each region of the planet. Considering that every celestial body has a hollow structure, during the entire history of Earth, due to recurrent shifts of poles, strata pressed under the weight of oceans was periodically lifted up to form mountains, and after an x-amount of time the same region could submerge under waters of the ocean again. Continuously re-forming itself, the planet was experiencing constant stretching and compressing of its surface until one day everything changed. The ejection of satellites caused a massive stretch of the entire surface of Earth creating an enormous difference of altitudes and crustal thickness accordingly. Continents that appeared after that event differ from each other by two factors: thickness of strata and its density.

We are who we are because of our habitat in a particular place on the planet. We have different climates, altitudes, magnetism, food and water. In other words, everything affects our bodies on the molecular level. The slowly flowing processes of mutation do not let us notice these changes, but they do exist. Considering every living thing on the body of our planet as a completely autonomous unit is absolutely impossible. Now, having said that, the utopian ideology of the Third Reich seems even more preposterous because even in the case of their victory in World War II and repopulation of the so-called Arian race over the entire planet, after centuries of following generations, the demographic situation of the planet would be exactly the same as before the invasion.

Getting back to the topic of stars and galaxies, there is another answer for the creation of life. Let us create a logical chain of facts. If satellites are the "children" of planets, planets are sparks from the stars and stars are one enormous firework of galaxies, then who are we?

Considering us and all living things as aliens to our universe, means to accept the existence of other universes where we could originate and later migrate from. But even that infinity does not answer the question of our appearance in the universe. The only answer is that we are the products of our universe, our galaxy and our solar system. All the flora and fauna we know and do not know is the product of our planet itself.

Any fresh grain placed in a sealed jar for a long time will eventually lead to the appearance of worms. They develop from the grain, they eat the grain and they die in it. Experiments with plants sealed in glass jars for years without any external watering is comparable to our planet. Sealed with its atmosphere, it created its own life forms. The expectation to find exactly the same life forms on other planets is absolutely useless.

"So God created man in His own image; in the image of God He created him..." But depictions of gods in ancient Sumer and Egypt do not look like us at all. Gods visiting us from different planets are aliens and cannot under any circumstances look like us because they are products of their own planetary environments.

By planting the same seeds all over the world, we will never get an identical harvest. If gods ever disseminated the same species all over the universe, then the number of cultivated races would be equal to the number of planets (habitats).

Now, knowing the history of our planet and our own, everything gets to its logical place. We have a hard time believing that once races of giants that once flourished on Earth are our direct ancestors. All the differences between us can be explained by our different habitats. Though living on the same planet they were flourishing in completely different environments.

Have you ever had the feeling that we do not belong here? Don't you think that we are too heavy for this planet? We are "too heavy" literally. Let us consider just one typical feature of our body: the spine. Every modern earthling has back problems. Supposedly the strongest part of our body is very vulnerable and has structural defects. If god(s) created us to fulfill their needs, then they could not make such a gross structural mistake. Mass production of defective robots just does not make any sense.

Another strange aspect is the length of our sleep. Today's technologies allow us to create robots, which are able to work day and night. If, according to Zecharia Sitchin, we were created as primitive workers, then eight working hours a day and then sixteen hours of resting would not satisfy the gods' requirements, especially in the face of the inevitable catastrophe of their own planet. Considering totally different environmental conditions on Earth in the past, people could have the ability to work much longer than us modern people. Gods were supposed to have created strong and enduring beings who could withstand and coexist with such monsters as dinosaurs. And as we know, there are no mentions of gods creating dinosaurs.

According to testimonies of people who have ever seen aliens, they describe them as green-skinned, large-eyed, small-nosed creatures. Whatever people associate them with does not matter; the main feature is that all of them have humanoid appearances. There is a great chance that all of us have the same ancestors, with only one difference: our habitats.

Many ancient myths tell us about alien reptile gods. Any prehistoric connection between humans and reptiles is uncertain. But there are some small details in our everyday life, which seem too suspicious to be coincidental. Have you ever seen pictures of the "reversed first toes" of dinosaurs or birds? There is nothing special. Then why is there such striking similarity with women's high heel shoes and why does it serve as a model of beauty for us? The same thing could be said about sunglasses. For some reason they also look more attractive for our eyes (according to many witnesses, aliens have large eyes).

What would be our conclusion if every person living in equatorial regions would have skis in their house? Of course, the first logical thought would be the former presence of snow in that region. Same comparison can be made with such a "useless" human organ as the appendix. From ancient times people cannot understand its true purpose. What we can assert here is that at some point of our history the appendix played a much more active role than it does now. Just the fact of its mutated existence indicates its larger role in the past. The only question is to find the perfect conditions for it to develop, which can be nearness to the sun, the absence of seasons, a higher magnetic field, higher levels of oxygen in the air and so on.

What conclusion we can make here is that we are who we are because of the environment we live in and our inheritance. So, besides the above mentioned **" We are what we eat (food wise)"** and **"We are what we absorb (information wise),"** we must add: **"We are who we are because of *where* we are."**

In our minds, the conception of our history is always connected with visualization of our prehistoric ancestors as some kind of monkey-looking humanoids. If Charles Darwin lived in the 21st Century, he would rename his "Theory of Evolution" into the "Theory of Mutation" because only major changes in environment can cause the mutation of any living creature in the universe. If biblical Adam and Eve were created on planet Earth, then it was completely different from today's planet and the visualization of biblical Adam and Eve as modern humans is completely erroneous. If, according to many ancient cultures, people originated on planet Mars (which is highly probable), then we can only guess about their original appearance.

Prehistoric Giants

Two distinctive discoveries have been made on the American continent in the 20th Century. In 1944, in the city of Acambaro, Mexico, (1865 metres above the present sea level), German immigrant Waldemar Julsrud discovered clay figurines of prehistoric animals. With the help of a local farmer, Waldemar Julsrud eventually collected over 32,000 figurines.

The discovery of the Ica Stones of Peru (at 400 metres above present sea level) also consists of quite a large collection. Thanks to Javier Cabrera (doctor by profession who devoted his life to collecting and studying these artifacts), the world got the opportunity to know about these precious findings. These andesite spherical stones also depict prehistoric dinosaurs. Doctor Javier Cabrera was obtaining those stones by purchasing them from a local farmer. But how were these hard stones engraved in the first place? Here again, all our research is based on the condition of these artifacts at the time of their discovery. A fossilized tree, for example, has quite a hard structure. After millennia of the fossilizing process, any carved wood would be considered as stone. Considering that before and right after the satellite ejection period the soil of our planet was much softer, then we can surely assert that at the time of engraving, Ica Stones were much softer than they are now.

(Pic. 42)

Besides the hardness of these stones, the most distinguishing feature of these artifacts is that there are people riding prehistoric animals depicted on them. But according to science, humans have never coexisted with dinosaurs at the same historical period of time. So, either dinosaurs were extinct not so long ago or humans existed on this planet for many, many thousands of years.

Curiously enough, nowadays we can purchase dinosaur related literature and toys in almost every bookstore, but with one big exception: there are no pictures or toys with humans riding, fighting or being eaten by them. It just does not even make sense for us to depict something like that because we were always told that humans have never coexisted with dinosaurs at the same historical time.

First of all, today's humans are far smaller than those depicted on these artifacts. As we know, rule number one for any sculptor or artist is compliance of proportions. Therefore, authors of these pretty complicated artifacts could not make such a gross mistake. They would not be able to create those models by the stories of others either; otherwise they would never have been able to depict them so precisely.

These precious artifacts give us an important clue about the life of all the prehistoric giants (giant animals and giant people). The Acambaro figurines were discovered at an altitude of 1,865 metres in North America scattered across the landscape with hundreds of fallen debris from ejected satellites. The discovery of one artifact that miraculously survived would not tell us much. But the impossibility of survival of such a large collection of fragile figurines under the massive bombardment of ejecting satellites indicates that these artifacts belong to a much later epoch when most volcanoes calmed down and allowed people to repopulate this area again.

The discovery of the Ica Stones of Peru at 400 metres above present sea level leads us to the same conclusion.

> 1. All Ica Stone depictions are made on andesite, which is volcanic rock.
> 2. All these perfectly rounded stones are debris ejected from Earth satellites.
> 3. The Ica province itself is located on a torn away slope of the Andes. Here again, the discovery of one artifact could mislead the determination of its origin. But the existence of such a large collection of Ica stones on formerly subterranean stratum indicates their belonging to a post ejection period.

During the moonless existence of Earth, due to a more intense centrifugal force, stronger magnetic field and higher levels of oxygen, our planet was inhabited by giants. During the First Great Catastrophe, entire populations of flora and fauna on Earth either vanished in the formation of ejecting satellites or were swept away by enormous tsunamis. Despite a drastic change of climate and its disastrous impact on all living things on Earth, some survived on the thin isthmus of the Andean Range. Flora and fauna have mutated and adapted to live in the new environment. Seemingly, extinct giants of the pre-ejection period eventually regained their population and continued to flourish until the Second Great

Catastrophe, where unceasing rain and the subsequent lowering of oxygen levels exterminated them once and for all.

Numerous discoveries of untouched giant burial sites in North America and around the globe indicate giants' miraculous survival after the ejection of satellites.

There are numerous maps at present with marked databases of giant remains. For example, the map of all giant remains ever found in North America shows a distinct similarity with the population density at present. This is a fact which leads to three possible answers:

1. Such evident correspondence of the maps can be due to a more intense construction in more dense areas of the present, and therefore higher chances to discover any findings beneath the ground.

2. It proves a relatively stable location of poles after the First Great Catastrophe. According to the frozen condition of animal remains found in Siberia and Alaska, after the ejection of satellites, one of the poles is always in the vicinity of the Arctic Sea.

3. Due to extreme air and water pollution right after the satellite-ejection period, correspondence of the arch-shaped density of the giant remains with exterior borders of a formerly monstrous ice cap of the pre-satellite-ejection period can indicate giants' post-catastrophic resettlement close to the fresh water supplies.

In conclusion, we can certainly assert that species of giant flora, fauna and humans had survived the First Great Catastrophe, and — according to these prehistoric depictions — people had been peacefully coexisting with dinosaurs, and the main cause of it was the abundance of vegetation. The absence of any findings, proving that civilization's more advanced achievements indicates that period between the First and the Second Great Catastrophes was:

- Either considerably short and civilization did not get enough time to develop.

- Or was distinctively fertile and the abundance of food and water eliminated the need to develop.

Unfortunately, due to the unpopularity of this topic at present, we have very little understanding of this civilization of giants, our far ancestors.

The Pyramids

From ancient times, our civilization has been trying to unravel the mystery of megalithic pyramids scattered around the globe. Their enormous size just fascinates our imagination as to their construction and purpose. According to our history, all Egyptian pyramids were built by pharaohs as gateways to eternity in their afterlife voyages. Acceptance of that theory automatically turns all the pyramids of the world into tombs. It seems like our ancient rulers had some kind of popular caprice to build unimaginably expensive tombs.

Let us imagine ancient rulers (not just one and not just on one continent) who, out of fun would decide to get into such grandiose construction without any pay back after its completion. Just as a reminder, pyramids were found in Central America, Africa, Asia and even in Antarctica (79°58'38.88" S, 81°57'37.68" W). The erection of such gigantic structures using all human and monetary resources of the country would be equal to self-destruction. It seems that all the ancient monarchs were seeking to add more headaches to their daily governing tasks. Despite the fact that the interiors of the pyramids are free from any carvings indicating their belonging to any civilization, we still try to link them to the ancient Egyptian civilization. But let us try to approach this enigma differently.

Knowing the history of our planet, we can try to link these marvellous structures to one of the three major epochs. So far, our civilization had discovered groups of pyramidal structures in North Africa, Central America and East Asia. If all the famous pyramids are located on three different continents, then we can assert the following:

1. If we suggest that the pyramids were built before the ejection of satellites and separation of continents accordingly, then despite their geographic closeness to each other before disintegration, no structure could possibly withstand such a global scale catastrophe without any damage, and the existence of ancient pyramids in Central America eliminates any possibility of their belonging to the pre-satellite-ejection period.

2. We could try to link the pyramids to our own quite superstitious civilization, but the existence of these structures on the continent of Antarctica definitely eliminates that possibility also, and just proves the recurrence of polar shifts.

3. The existence of the similarly designed pyramids on four continents of our planet indicates centralized governance during their construction, and that could be possible to achieve only by a highly developed worldwide civilization or by some (alien to our planet) civilization that could easily travel around the globe.

In order to determine the origin of the pyramids, we have to try to understand their purpose, and one of the best places to go is, of course, the Giza Plateau. The interior structure of the Great Pyramid shows scrupulous building techniques on its upper layers and a very rough and poor treatment in its basement. That fact can point at two things: the importance of the pyramid's upper levels over its basement or, better yet, the impossibility of conducting such engineering perfectionism in the pyramid's lower levels.

According to many archaeologists, there is a great difference between the exterior perfectionism of the Great Pyramid and the layers behind. Beneath the ordered exterior masonry of the pyramid, there are tons of uncut and irregularly shaped blocks, which are filling the gaps between exterior walls of the pyramid and its core. That is correct, its core.

The only explanation for such a discrepancy is that the construction of the pyramid had never even started from its basement. Considering that the entire African continent, the Arabian Peninsula and the surrounding

Giza Plateau regions are scattered with small volcanoes, then after heavenly bombardment the entire region should be covered by steam and smoke.

But why would anybody try to build anything in such a harsh environment? Nowadays our civilization uses hot underground waters to heat their houses and extract energy from them. A good example is the still hot island of Iceland. Taking into consideration that all the pyramids of Giza are located along the Nile River, then all the underground compartments of the pyramids had to be filled with water — and if we are talking about volcanoes — then they were filled with boiling water.

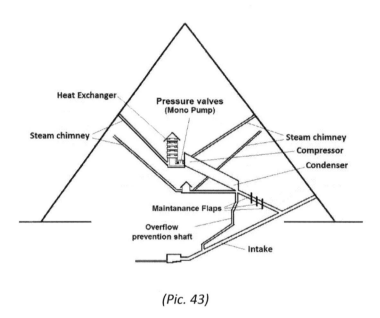

(Pic. 43)

Here is an image of the Great Pyramid's interior with a completely different vision of its purpose *(Pic. 43)*. Taking into consideration that at the time of its construction the "subterranean chamber" was extremely hot, it would be logical to provide a temporary escape channel (descending passage) dug right into the chamber to keep deadly steam away from the construction above. In order to continue safe masonry, three flaps were installed (Girdle Stones) to prevent hot steam from escaping into the main compartments during construction. After completing the construction, the "Grand Opening" of the pyramid would be the opening of all three flaps.

The pyramid would work as follows: travelling from underground, steam passes through "pressure controlling flaps" into the "condenser" ("Grand gallery") where it cools down on the tilted walls and flows down, depositing its heavier particles in the pockets located along the entire length of the tilted gallery. Meanwhile, the main part of the condensed water flows to an "elution column" (intersection of the "Grand gallery," "Ascending passage," "Horizontal passage" and "Well shaft"), where its heavier particles fill the dropped down section of the "Horizontal passage" ("Jubilee passage") to conduct "gravity concentration" where its lightest particles flow away through the "Well shaft" preventing the overfilling of the chamber.

Constantly arriving steam accumulates at the top of the grand Gallery, creating enormous pressure on its uppermost part. Reached to its maximum value, high pressure would periodically open pressure valves of the antechamber to supply highly pressurized steam into the tank (the King's chamber) to conduct pressure oxidation. To protect the chamber from extremely hot steam, the top of the chamber is made of layers of stone blocks with air pockets in between; the perfect way to prevent the chamber from overheating (we use the same heat sinks to cool down electronic parts). The existence of such a massive cooling mechanism above the "King's chamber" proves its structural purpose, which allows conducting high temperature and high-pressure reactions. The "Pharaoh's sarcophagus" could be filled with liquid mercury to absorb all the fine gold particles from the steam pushed into the chamber to form amalgam. Not coincidentally, the earliest proof of mercury's usage in ancient Egypt was found in an ancient tomb at Kurna dated back to 1,600 BC.

Undeniable proof of a highly efficient process of gold extraction via mercury hot springs is one of the most productive gold mines of present times — the Hishikari mine in Japan. Its productivity and uniqueness is based on the elution of gold from active volcanic hot springs pouring from the still fresh wounds of our planet.

Now, knowing the work principle of the Great Pyramid we can notice two strange features:

1. The volume of intake exceeds the volume of outflow by almost eight times, a direct indication of additional, undiscovered air shafts. (Imagine building a hydro power station with lesser outflow capability than inflow. It would be destined to disaster). Apparently, with the volcano's gradual cooling, the air shafts had to be gradually sealed off to sustain the necessary air pressure.

2. The liquid that flowed into the "Queen's Chamber" had to have (still undiscovered) an out-flowing channel(s).

Considering that the interior of the pyramid had been built to sustain enormous pressure, then all the hidden passages should have doors which open towards the pressurized interior because maintenance of such a plant would require access to its every chamber.

Now let us get to the exterior design of the structure where the main task of the builders was to tame the steaming wounds of the planet with heavy blocks to elute gold and to prevent it from exploding.

Building a triangle, pentagon or a more sided pyramid would require high precision in the masonry of blocks. A shape of cone structure would require achievement of perfect joints between the blocks where seamless joints of exterior blocks were a crucial requirement of the structure. The escape of the steam through the gaps would bring the entire project to a halt. That is the exact reason for such perfect joints between the blocks. Here we can make the conclusion that a square-based structure is the most effective and proficient.

The other distinctive feature of the ancient pyramids is their completeness. In other words, if we compare our civilization's odd looking plants with the geometrically perfect shape of the pyramids, we cannot even imagine their industrial purpose, but the last cube on the top locks all the sides of the pyramid together. That is the exact reason for the completeness of the ancient pyramids. Besides its main purpose to withstand the pressure from beneath, the upper part of the pyramid serves as a roof. In fact, by comparing our civilization's walls and ceiling structures with the two-in-one structure of pyramids, we can see their obvious advantage. Any manmade structure on earth should be protected from weather and, apparently, the square-based pyramid is the most effective shape to withstand earthquakes, heavy rains and strong winds. By the way, earthquakes in the period right after satellite ejection were much more powerful.

The purpose of the concave shape of the Great Pyramid's sides is to achieve additional structural strength to withstand internal pressure. Flat surfaces closest to the core are the most vulnerable parts of the pyramid. This concave feature of the pyramid increases its stability in times because in order for internal pressure to break through the closest wall, it has to push the pyramid's heaviest parts (its corners) away. It is always easier to break a straight stick than a convex one.

Now, by assembling a centuries-old puzzle, we get to the next great discovery. Presently all the Giza pyramids are located approximately 40 metres above present water levels of the Nile River, which means that back in time, waters of the Nile River were much higher. The existence of such a massive volume of water could be possible:

1. If the entire northern part of the Africa had a tropical climate. If we suggest that the geographic pole at that time had a different location, then the basin of Lake Victoria filled to the top could be the potential source of such a massive water supply.

2. If the sea level of that particular region was much higher than it is today.

Proof of much higher water levels lays in the discoveries of so many boats in the pyramids' vicinity. The discovery of one Funerary Boat of Khufu could possibly point to its ritual purpose. But discoveries of more than ten boats of various sizes indicate the existence of a body of water at that particular time and nothing else. It is like making an analogical assumption that all the ships in the desert of the former Aral Sea are burial sarcophaguses of almighty kings of antiquity. Nonsense, isn't it?

Discoveries of boats in so-called pits indicate their existence in a completely different climate. If we assume a closer proximity to a geographic pole than at present, the Giza Plateau would experience a change of seasons. If we imagine an enormous harbour with hundreds of boats tied up in its piers, then the approaching fall would force seamen to protect their boats from freezing, especially in the fresh waters of the river. Here again, I would like to traverse us to that time distant from our epoch at the ejection of the satellites and the following bombardment of our globe that led to the occurrences of continuous acid rains, and the need to protect the wooden boats would be quite necessary. Dismantling the boat in order to minimize its size and store it beneath a stone canopy seems very plausible. If we are impressed by the sizes of the blocks covering the pits, then they are no bigger than the blocks used in the construction of the pyramids themselves. We should not forget here that in order for a pyramid to work as a gold producing plant it had to be located on or near the water. So, there is a big chance that all these boats belonged to the staff members who were maintaining the pyramid.

Now, knowing the real purpose of the Egyptian pyramids, let us traverse to the pyramids of Central America and try to link them together. The first similarity that comes to mind between them is their pyramidal structure. But there is one more and most important similarity in their design, which links them together in purpose. It is the step-like structure of the Central American pyramids and step-like structure of the grand galleries of the Egyptian pyramids. Identical to the Grand gallery of the Great Pyramid, the pyramids of Central America were serving as gold eluting factories with the only difference in their design. In both locations, the process of gold elution consisted of the condensation of steam and the following extraction of gold.

In 2013, an ancient pyramid in Belize was destroyed and construction workers decided to use the pyramid materials to build a road *(Pic. 44)*.

The picture of the remaining untouched part of the pyramid shows very primitive masonry with only one hole in the centre. The absence of perfectionism in the masonry indicates that it was not necessary to the construction, and the large vertical hole in the centre of the pyramid served as a steam passage. The same shafts exist on all the megalithic structures around the globe. The famous Tomb of the Red Queen in Palenque was also installed on the top of a deep shaft. And again, the archaeological mixture of the two epochs was wrongfully interpreted as a "shaft connecting the dead with the underworld."

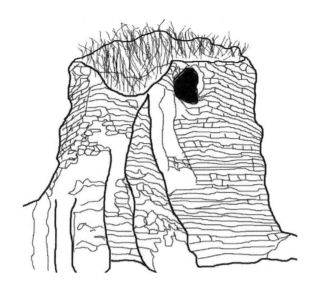

(Pic. 44)

In contrast to the Giza pyramids which had enclosed and more complex cycles, the pyramids of Central America had much simpler structures. Most of the Central American pyramids have house-like structures on their tops and door-like openings facing the steps on each side of the pyramid. Water condensed on the flat roof was travelling down the steps, depositing its heavy particles on the slopes of the pyramid. It would be logical to have some kind of shifting mechanism on the top of the pyramid allowing the alternate distribution of condensed water on each side to let the slopes of the pyramid cool down in order to collect the deposits.

In 2015, Mexican archaeologists announced the discovery of an enormous sinkhole beneath the famous Temple of Kukulkan in the Chichen Itza complex. As we know, the entire Central American continent is scattered with sinkholes as a result of an enormous shift and "mudslide" of continents. Such a painful process as an ejection of satellites was accompanied by the emission of immense heat. As in our bodies, where blood is the healer of the wounds, the cooling down of the planet's wounds was carried by a liquid also: water. Here we can assert that after the ejection of satellites all the subterranean channels were carrying boiled water. That is the exact reason of such a chaotic distribution of all the pyramids on our globe. They were built on sinkholes to extract gold from boiling water. The recent discovery of liquid mercury in the chamber below the Pyramid of the Feathered Serpent (one of the pyramids of Teotihuacan, Mexico) just proved the gold extracting purpose of all ancient pyramids.

The next closely constructed pyramids are located in China. We do not know much about their interiors but knowing the purpose of the Egyptian and Central American pyramids we can assume that the pyramids of China have the same enclosed elution structure. The discovery of the Baigong Pipes and a mysterious pyramid in the Qinghai Province of China points to the existence of hot water in a nearby lake in the past.

Now, let us scrutinize the direction of the world pyramids. All the Egyptian pyramids face towards the present north and south directions. The pyramids of China and Central America, on the other, hand face directions completely different from one another. Now, let us think logically. If every "tectonic plate" on our planet shifts 2 to 11 cm annually and has its own individual trajectory, then the megalithic structures of antiquity cannot possibly face the directions they were originally built in. Therefore, the orientation of the Egyptian pyramids to the present north-south directions is nothing more than pure coincidence. We constantly admire the precise north-south direction of the Egyptian pyramids while simultaneously studying the movements of tectonic plates without linking the structures located on them to the continents themselves.

After rediscovering the real purpose of the ancient pyramids, we can make the assertion that out of three major pyramid locations on Earth (Egypt, Central America and China) the most truthful description of their builders was preserved by the Indian Tribes of Meso-America who have never claimed that the pyramids were built by the almighty rulers of antiquity. In contrast to that, by proclaiming the pyramids as burial tombs, the ancient rulers of Egypt and China rewrote the history and misled the future civilizations, and therefore any archaeological discovery should always allow the possibility of being discovered by earlier explorers.

Much like animals that mark their territory by spreading their personal odours, we people like to write our names on every writable surface (be it a cliff, cave or even public washroom). We do not know much about Khufu, who ruled ancient Egypt many thousands of years ago. He pocketed the erection of the pyramids and therefore misled all the civilizations that followed.

Our obsessed search for pharaohs' treasures hidden in the pyramids has only one true coincidental factor — it was full of gold indeed. The last extracted gold was taken from there by the maintenance staff left by the gods to maintain the constantly shrinking pressure of the underground steam. The sealing of every air shaft indicates the pyramids' production until the last drop of water. It would not be reasonable for the gods to stay any longer on our planet and it would be more logical to leave crews of faithful earthlings to collect and store the pyramids' last extracted gold. That was the reason for the gods' last promise to return soon.

Dolmens

Now, by revealing the purpose of the ancient pyramids we can easily explain the purpose of less complex megalithic structures — dolmens *(Pic. 45)*.

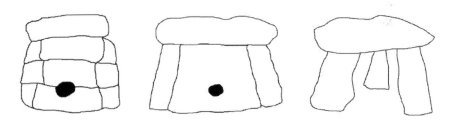

(Pic. 45)

Megalithic structures with quite primitive architectural design scattered around the globe also have one — and only one — purpose: to extract gold from the fresh, steaming wounds of our planet. Of course, they would not be as productive as the pyramids but the desperation of the gods to save their own planet from the inevitable weakening of its atmosphere forced them to collect every gold-containing drop of water.

Here again, the main principle of these structures is the extraction of heavy particles from water for later extraction of gold. Considering Dolmen's typical architectural feature with a round hole on its side, it would be logical to use some kind of liquid-soaked mercury filter to absorb all the golden particles. These filters could be periodically refreshed and, with the gradual weakening of the steam pressure, they would be taken away or rot away forever. Besides their most common structural features, some dolmens have slightly complex designs with obvious gold-extracting features.

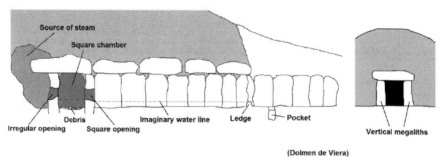

(Pic. 46)

Unlike ordinary dolmens, the "Dolmen of Viera" *(Pic. 46)*, located on the Iberian Peninsula, has a slightly different design and gold-extracting method. All the walls in this complex are vertical. The square chamber located in the end of the corridor consists of four megalith walls. The front and back megaliths have openings in the centre. Remarkably, the front megalith has an almost perfect square opening; meanwhile, the block on the back has a very odd-shaped cut-out.

Apparently, during its construction the back megalith served as a temporary shield to allow the safe completion of the project. A small explosion on the back wall would mark the beginning of the operation. In contrast to the fairly smooth floor of the corridor, the floor of the square chamber is scattered with stone debris that are fused together. This is an indication of its inaccessibility after the mini-explosion due to the pouring of steam through the opening.

Located in the same area, the Dolmen of Menga and the Cave of Romeral *(Pic. 47)* have more prolonged corridors leading to the source of steam. In the Dolmen of Menga the source was reached through a 20-metre-deep shaft. In the Cave of Romeral, a megalith block on the floor was laid to guide upcoming steam from beneath towards the second dome-like chamber. Besides stability purposes, the main feature inherent to all megaliths is their tilted corridor walls. As well as in the grand gallery of the Egyptian pyramids, tilted walls are required so condensed water will drip better.

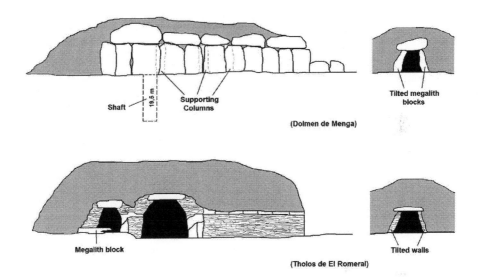

(Pic. 47)

Knowing the purpose of this plant we notice the obvious resemblance to the interior structure of the Great Pyramid of Giza. The square chamber in the Dolmen of Viera corresponds with the King's chamber of the pyramid. Both chambers are designed to withstand high pressure reactions. Steam pouring from the chamber would fill the corridor with condensed water. Considering that the height of the square opening is slightly higher than the ledge located close to the entrance, water would spill over from the ledge, leaving its heavier particles behind. The existence of the ledge indicates the necessary level of water, and the pocket located next to it could serve as an additional gravity concentration tub *(Pic. 46)*.

Discoveries of the Burial Mounds in Europe and Asia dazzled robbers and misled archaeologists about the original purpose of these structures.

Of course, any well-preserved archaeological site assumes its originality, but little did the archaeologists know that the famous discoveries of the Scythian tombs belonged to two completely different epochs, and the almighty rulers buried in these "tombs" had absolutely nothing to do with the construction of these dome-shaped structures.

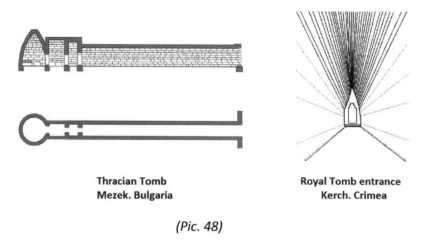

Thracian Tomb
Mezek. Bulgaria

Royal Tomb entrance
Kerch. Crimea

(Pic. 48)

Identical to the famous dolmens of Spain, the Thracian Tomb in Mezek, the Scythian mound of "Tolstaya Mogila" near Dnepropetrovsk and all the tombs in Crimea — or any other region of the world — are in fact dolmens. They were built by a race of giants before the Great Flood and way, way before the appearance of the Thracian, Scythian or any other great civilization of antiquity. In fact, these gold-eluting structures were discovered by our ancestors thousands of years before us and were used as burial tombs for their leaders. Little did these people know that transforming these ancient structures into mounds would mislead the entire understanding and history of the civilizations to come.

Now, knowing the purpose of these structures we can generalize our conclusions. Just the fact of <u>worldwide distribution of such architecturally primitive structures</u> indicates their highly intelligent builders because primitive societies of ancient giants would not be able to erect similar architectural structures around the globe. Being built by a race of giants, these structures would be absolutely useless for them to use as homes, shelters or even tombs because no giant body would possibly fit inside. The discoveries of human and animal remains inside of many megaliths worldwide do not reveal the purpose of these structures. Being built by a vanished civilization of giants, postdiluvian Earth became populated by smaller stature humans. The following superstitious civilizations unwillingly transformed all the megaliths into tombs.

The worldwide distribution of similarly designed structures can be possible only in the case of centralized governance capable of <u>flying</u> around the globe in order to collect extracted amalgam. Gods who arrived to our planet in search of gold would not be able to bring all the heavy industry of their civilization with them. To build the same industry on a new planet would be impossible and useless at the same time. We should not forget that gods were sent to every corner of our galaxy to mine for gold in order to save their own planet, and one of the main problems they had was that they were running out of time. That is why they came up with the fastest and most efficient methods to extract gold from geysers.

Here we get to the conclusion that all the dolmens are not ritual structures inherent to the architectural style and traditions of the local tribes. They were built by a race of gods who, like bees, flew around our globe in search of precious metals. Architectural designs of any megalithic structures on earth were dictated by power of emitting from beneath steam. The existence of primitive dolmens indicates weak sources of steam and the complexity of others indicates their purpose to handle larger volumes of steam. Generalizing the above mentioned structures, we can surely assert that erections of such complex structures as pyramids were dictated by enormous power, and the chemical richness of belching geysers at the Giza Plateau, Central America and south-east of Asia.

One of the main characteristic common to all these structures is their unusual energy. As per evidence of many grotto researchers, people who have been inside had experienced very strange and mixed feelings. Considering that all the prehistoric megaliths were built on the tops of steam shafts, after the gradual cooling of Earth's wounds, the remaining grottos became arteries for transmitting electromagnetic waves, thus connecting deeper layers of the globe with the outer surface of Earth. Electromagnetic waves of deeper altitudes travelling through the pores reach the surface of Earth and the frequency alien to Earth's surface has inadequate effects on every living thing.

Here is a simple example. A tourist travelling around the globe experiences the influence of diverse electromagnetic waves. The main cause for such diversity is the different thicknesses of the crust in different regions of the globe. Waves beaming from pores different in frequency are forming one unified and global background, and the diversity of frequencies does not cause any harm to the traveller. In contrast to a diluted bundle, passage through underground shafts means being right in the epicenter of an electromagnetic stream of one particular frequency. As we know, some frequencies cause the appearance of sudden panic and that is the reason for such a bad reputation of many caves around the globe.

Therefore, we can make a simple assertion: if every megalithic structure of the post-satellite-ejection period was built to extract gold, then the source of this "bad energy" is hidden inside of our planet and not inside of these structures.

Numerous discoveries of subterranean pyramids in many regions of our planet prove that they were built in the antediluvian period. In a matter of days (the biblical forty days and forty nights), megaliths located in mountainous regions were covered by dirt and debris brought on by the Great Deluge. All the great pyramids we know today are located on plain landscapes and that is the main factor of their survival to our days. All the structures located at the foothills of mountains were simply buried under massive mudslides.

The main misleading ideology of our civilization is that we were assured of our primary origin on this planet. As soon as we admit the existence of many civilizations preceding us, every archaeological discovery will be considered as layers of many historical epochs. In other words, whatever we discover today might have been discovered before us by our preceding civilizations.

Now, by knowing many answers, we can certainly assert that no dolmens, pyramids or any other megalithic structures on Earth were built as a tomb or device for the after-death journey into eternity. Under the guidance of the gods and built by a race of giants, megaliths constantly inspire civilizations that follow to associate these mysterious structures as an apparatus for communicating with gods. Of course, it is a superstition, but I guess some sixth sense perceives the fact that these megaliths were touched by the gods — literally.

Boulders

After the exploration of prehistoric megaliths we must traverse to the Indian peninsula. The Sigiriya and Pidurangala Mountains are two extinct volcanoes positioned one kilometre apart from each other, and located in the heart of the island of Ceylon (present-day Sri Lanka). The main archaeological significance of the Sigiriya rock starts from the two claw-type megaliths at its base *(Pic. 49)*.

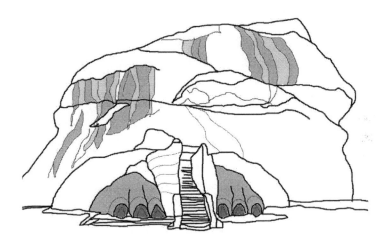

(Pic. 49)

Moulded or carved from the rock itself, claws make the first and very majestic impression but the main fascination awaits us on the top of this 200-metre-high mountain. The terraces carved on the top of the mountain are said to be made by an ancient Sri Lankan king who ruled the country in the 5th Century AD.

This historical narrative would perfectly fit into our civilization's biography if there was no evidence pointing to the antediluvian origin of these structures.

- If the author of these monolith claws decided to carve some animal body from the rock, then plans would include the creation of a torso also. The absence of a mountainous body right above the claws indicates the planning of future build-up on the surface of the mountain using some kind of compound unknown to us.
- Numerous multicolour stripes on mountain slopes indicate their volcanically active past, which would definitely attract the gods' attention, as apparently it did.
- Numerous terraces and pools on the mountain top are made of megaliths (present people are way too small to move such weights).
- During rainy seasons, numerous perforated hatches in the water gardens located at the foot of the mountain periodically disgorge water into the fenced reservoirs.
- The last and most compelling evidence of the gods' creation are enormous rocks located on the tops of these and many other highland mountains around the globe *(Pic. 50)*.

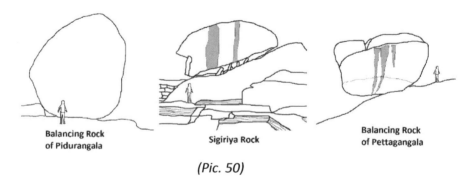

Balancing Rock
of Pidurangala

Sigiriya Rock

Balancing Rock
of Pettagangala

(Pic. 50)

In the pictures above we can see enormous boulders installed on the tops of the mountains. These megaliths, some of them perfectly balanced with vertical pillars supporting them from underneath, also served as caps to liquefy the steam gushing from the throat of the volcano. Here we can detect soot traces on the sides of these boulders and their obvious similarity with dolmens scattered around the globe. In contrast to pyramids and dolmens (where the flat landscape was allowing the building of more complex structures), builders of these gold-eluting plants were limited by high altitude and unevenness of the landscape.

We should not forget here that these megaliths were not erected in favorable construction conditions. If, at present, in order to build any high-rise, we are able to flatten any landscape, builders of these prehistoric plants did not have such freedom to do so because the entire landscape of this region was scattered with numerous funnels in the ground constantly or periodically belching superhot steam. The capability of builders to install such monstrous boulders on the tops of these mountains is impressive enough because modern human beings would not be able to roll such enormous and heavy objects to the top of a hill, not to mention the complete uselessness of such an act at present due to the absence of steam.

Identical to the above-mentioned boulders of Sri Lanka, there are numerous boulders in Burma. In contrast to the balancing rocks of Sri Lanka, the Balancing Rock of Kyaiktiyo or Golden Rock of Burma have one very curious feature. Along with the miraculous method of its installation, there is also the dazzling brightness of its golden surface. Traditionally, all the pilgrims who visit the rock affix golden leaves on its surface as a sign of respect and veneration to Buddha. Years of this gold-plating ritual have turned the rock into one enormous, seemingly solid nugget of gold. It is quite symbolic. Playing one of the main roles in the elution of gold, this tool was gold-plated by future generations of humans.

The balanced boulders of Sri Lanka, Burma and India are not unique and can be seen in many regions of our planet. Their most common feature lies beneath them where every one of them covers a deep shaft connecting the surface of Earth with its presently healed wounds. Here we should not confuse the naturally gradual eroding sedimentary rocks with the rounded shapes of artificially balanced rocks. The rounded shapes of the boulders were needed for easier transportation and ascending of the megaliths to the top. In fact, the size and shape of these boulders can tell us a lot about the stature and capabilities of its installers.

We must consider two possible methods for installing these boulders: either they were rolled up the hill from the base or they were moulded right on the top of the hot mountains, and later rolled towards the gushing holes. Independently of their origin, both methods point to a single fact: the boulders were rolled towards their final destination, and this fact eliminates myths of rock levitations in antiquity because to maximize the surface of the cap it would be logical to use flat shape rather than spherical. Installation of such unstable boulders at the tops of the mountains was not done for fun; it was dictated by the absence of other more appropriate tools to stop the steam from evaporating.

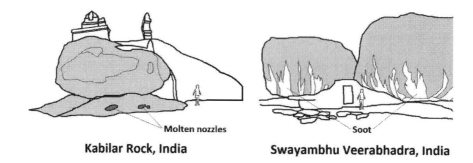

Kabilar Rock, India   Swayambhu Veerabhadra, India

(Pic. 51)

In the picture above *(Pic. 51)* we can see two more sites of megalithic boulders located in India. Realizing their divine origin, our ancestors turned these sites into holy places.

What caught my attention here was the molten nozzles at the base of the Kabilar Rock, and traces of soot on the Swayambhu Veerabhadra boulders — an obvious proof that very hot processes flowed here in the past.

Ancient structures

Now, after examining comparably primitive structures, let us consider more complex gold-eluting plants of antiquity. Our first stop should undoubtedly start on the Indian subcontinent, which is, by the way, a very strange name for a peninsula.

The diversity of temples here just strikes our imagination. In order to start, we must consider one important attribute of Hinduism: the symbol of energy and potency — Shiva Lingam. These artifacts had been installed in almost every Hindu temple *(Pic. 52)*.

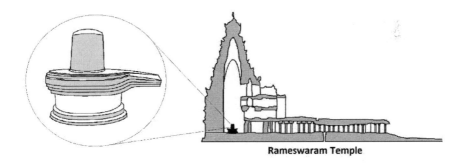

Rameswaram Temple

(Pic. 52)

Shiva Lingams consist of three separate parts that allow for easy assembling into one unit *(Pic. 53)*. Despite their quite simple form, these structures have two typical features:

1.   The elevated rim around the upper plate has an arrowhead neck for the collection and flowing of the liquid into one direction (Figure 3).
2.   The lingam itself (Figure 1) is made of pure (99.8%) solidified mercury, which is impossible to achieve with modern chemistry.

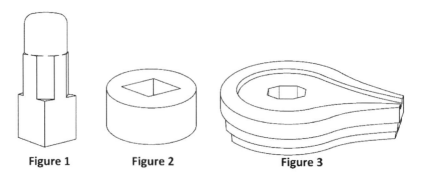

Figure 1 Figure 2 Figure 3

(Pic. 53)

Now, let us try to unravel their purpose. The rim around the upper plate is to guide the liquid to flow in one direction. The liquid itself should drip or pour from above, but considering that the majority of Shiva Lingams are installed in the centres of temples with their gradually narrow dome-shaped openings, we can make the assertion that these devices were not created to receive water from above but — on the contrary — were built to withstand the water pressure from beneath.

Considering that the majority of original Shiva Lingams are partially submerged into subterranean waters we can create an exact scheme of their work. The opening of vapour gushing from the ground was covered by a ring with a square bore (Figure 2). Inserted into the square opening (Figure 1) was thereafter capped by the rimmed plate (Figure 3). The usage of such a heavy and enigmatic element as mercury was necessary to hold the steam pressure, filter it and catch the gold particles. Very often, the nosing of the plate is installed right above the pocket in the ground to collect all the heavy particles of water.

The pin inserted into the stream of water makes water spray in every direction, forming a flower-shaped fountain. Dome-shaped roofs around the lingams are built to collect the water. Enclosing the domes completely would be unthinkable because the power of the fountain could definitely vary. A small opening on the roof was necessary and wasteful at the same time, which is why many roofs are capped by round stones resembling our present-day roof fans that have ribs on their sides.

Our next stop is the village of Lepakshi with its marvellous temples and famous Shiva Lingam installed right next to an enormous rock with the cobra-shaped screen *(Pic. 54)*. Despite skillfully carved depictions of cobra heads on the rock and a snake's body around the lingam, the main feature of this structure is the size of the rock itself. The screen on the rock was needed to bounce the water gushing from the lingam, and such a heavy weight of the rock was needed to withstand that enormous pressure.

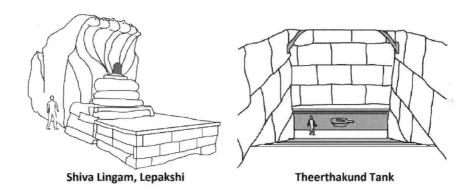

Shiva Lingam, Lepakshi   **Theerthakund Tank**

(Pic. 54)

The picture on the right *(Pic. 54)* depicts the subterranean tank of Theerthakund with a submerged into the water lingam. What impresses the most is the size of the stone blocks. These megalithic blocks can be compared with the stone blocks of the Great Pyramids or the megalithic walls of South America, but the spiritual importance of the lingam overshadows the obvious: these megaliths could not be built by our race. They were built by a race of giants, and giants used to live on our planet before and right after the First Great Catastrophe. Considering that the Indian subcontinent is the result of a massive landslide followed by the ejection of satellites, we can make an assertion that all the megalithic structures on that peninsula belong to the right-after-satellite-ejection period, and all the marvelous carvings on these megaliths were done by subsequent generations.

In fact, complex of structures at Lepakshi can unravel many mysteries of our civilization. Along with admirably carved sculptures, we must pay attention to the huge monolith block sitting in the middle of that complex. Untouched by masons, part of the "Hooded serpent" indicates that it was there long before people's intention of carving it. Luckily for us, the masons of this monolith rock came up with carving the front of it without touching its main body. These counterweight monolith blocks are scattered around the globe with one intention only: they had to withstand gushing water pressure from beneath. And located 200 meters away from the main complex, the sculpture of Nandi bull was carved from one of these monolith rocks.

Let us look at one of the most famous and "mysterious" sculptures on earth, the Great Sphinx of Giza. Besides the absence of any proportions in it, the main feature we should pay attention to is its similarity with many sculptures of Lepakshi. As we can see here, it also was carved from one solid megalith block. On the picture below *(Pic. 55)* we can clearly see the main features of these rocks, which motivated the masons to sculpt their masterpieces.

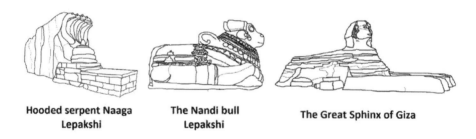

Hooded serpent Naaga
Lepakshi

The Nandi bull
Lepakshi

The Great Sphinx of Giza

(Pic. 55)

All these rectangular blocks had protruding lumps to increase the surface area for withstanding the pressure screen. What else could ancient masons sculpt if the look of these rocks resembles something they were seeing every day? It would be lying on the ground animal. Depicted by mason bull became sacred animal on Indian peninsula and sculpted by ancient mason creature led to idolizing the sphinx in Egypt.

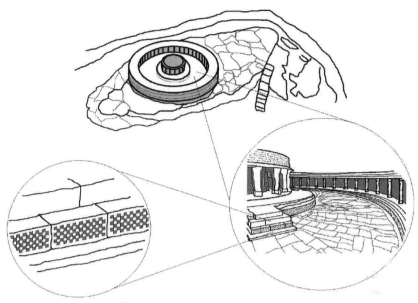

Chausath Yogini Temple, Morena

(Pic. 56)

Along with the dome-shaped structures scattered on the entire Indian subcontinent is the Chausath Yogini Temple, Morena *(Pic. 56)*, which has quite a different architectural design. Located on the top of the hill, it consists of a circular rim with a round temple in its centre. Of course, it could have been built as a shrine, but a very odd thing catches one's eye: the continuous row of perforated stones on the entire inner perimeter of the circular passageway between the two buildings.

To build such a massive drainage system to accommodate just rainwater would be quite pointless. If we assume that at the time of construction this region was receiving massive rains, then what is the point of living there? It was built to accommodate a massive amount of water and that amount could only be supplied by geysers.

The next, seemingly ordinary temple is located right next to the Shaiva Temple of Avudaiyakoil (Tirupperunturai). But what is not ordinary here is its ceiling. Made of granite, the Ceiling of the Kanga *(Pic. 57)* is an absolutely stunning piece of art. This unique ceiling has been carved (or moulded) in the form of joists riveted together, flat bars and ropes. The purpose of this complex ceiling decor is quite obvious: to cool down super-hot steam. The larger the area of exposed surface, the better the cooling down ratio. We use the same principles to cool down our electronic components.

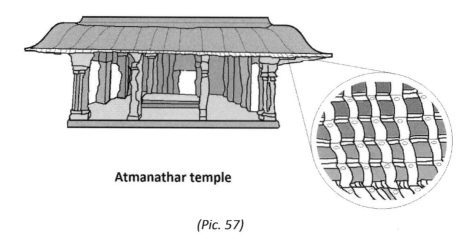

Atmanathar temple

(Pic. 57)

For people living in this era, the creation of something like that would take an unimaginably long period of time, not to mention the possibility of errors during the carving process because one stone chip would ruin the entire beauty of this masterpiece. Here we can make two conclusions: either ancient people were errorless (which is very doubtful) or this knitting was done with a fairly elastic material. We should not forget that this global "gold rush" took place right after the satellite ejection period when the entire crust of the earth was much, much softer than it is now; therefore, all the structures of the antediluvian period were formed and moulded from flexible and much softer granite.

The picture below *(Pic. 58)* depicts two temples located thousands of kilometres away from each other but built to serve the same purpose — the transformation of hot steam into water and its following filtration.

The Prambanan Temple complex is located in Indonesia and has a very unique architectural design consisting of six main temples surrounded by 224 much smaller temples. But the main features we need to discuss here is the deep well located underneath the altar and the two empty rooms

162

above the main chamber. The well underneath former Shiva Lingam was the steam supply and the empty rooms above served as an additional method of cooling. As we can see, the design of the underground well has a striking similarity with the well of Dolmen De Menga *(Pic. 47)* or the pyramids of Central America.

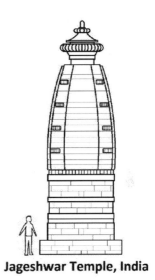

Prambanan Temple, Indonesia   **Jageshwar Temple, India**

(Pic. 58)

The picture on the right *(Pic. 58)* depicts one of the temples of Jageshwar. Located at the foot of the Himalayas, it is a cluster of structures consisting of 124 temples varying in size. In contrast to the Prambanan Temple of Indonesia, these temples do not have deep wells underneath (they just have not been discovered yet) and have served the same purpose of cooling. The cooling down process here was implemented through round, ribbed stones inserted into the body of the temple, as well as with the large one mounted on the top. Water flowing into a nearby reservoir would be treated to extract gold particles.

After one sees the stepwells of India, the water reservoir of Jageshwar will seem like a small pond. Scattered on the Indian Peninsula, stepwells have quite unique architectural designs *(Pic. 59)*. Consisting of hundreds of steps, these upside-down pyramids have just one purpose: to collect water for further processing to extract gold. Despite the diversity of design, most of the stepwells have one inherent feature — two horns sticking out that are located on the only flat side of the stepwell. For us modern people, it resembles supports for a collapsed balcony. We must admit here that to build a balcony above the ditch-water pond would not be very attractive. If we propose the purpose of these horns to be electroplating electrodes then it would be quite a logical method to extract gold, and one of the main arguments is the colouration of these electrodes. Some of them have a greenish tint (possibly oxidized copper) and some of them are black (cast iron).

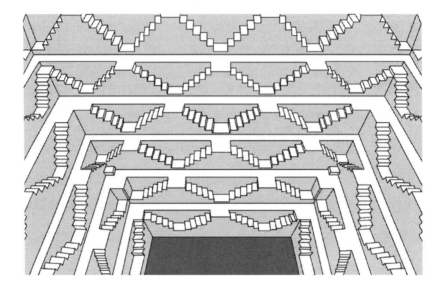

(Pic. 59)

If, in order to tame the geysers gushing from underneath the earth's surface, the gods were building temples, then the water supply of step-wells remains unclear. To suggest that water was gushing from underneath the well does not seem plausible because of the impossibility of conducting any excavation, and following masonry of the crater in such an extremely hot environment. Considering that the majority of the stepwells on the Indian Peninsula are spread in rows from north-east to south-west *(Pic. 60)*, we can make an assertion that all these structures were built on a very rich gold vein. The only possibility to build such architecturally challenging structures would be the temporary diversion of the gold-rich riverbeds, just like we build our power plants at present.

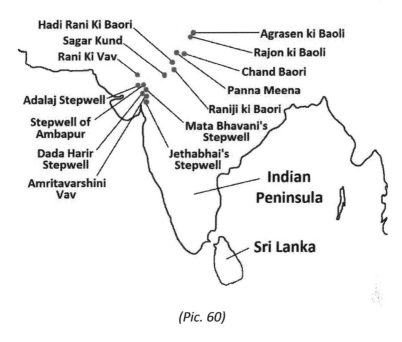

Hadi Rani Ki Baori
Sagar Kund
Rani Ki Vav
Adalaj Stepwell
Stepwell of Ambapur
Dada Harir Stepwell
Amritavarshini Vav

Agrasen ki Baoli
Rajon ki Baoli
Chand Baori
Panna Meena
Raniji ki Baori
Mata Bhavani's Stepwell
Jethabhai's Stepwell

Indian Peninsula

Sri Lanka

(Pic. 60)

Spending such enormous resources to build such great water reservoirs would not make sense if the water stayed there forever. Therefore, the purpose of such a grandiose construction was to make water flow through its centre. If we are to suggest that water flowing through the centre of the well could easily generate the necessary electroplating voltage, then oppositely charged horns in the well could possibly attract gold particles.

Compared to the rest of the stepwells of India, the stepwell of Chand Baori *(Pic. 61)* has quite a compelling design. Along with its massive size, decorated facade and existence of horns, the lower part of the well has ribbed tires installed right at the tops of two reminding mini-temple structures located on each side of the horns. The same tire-like stones had been used on other archaeological sites to prevent the stone from overheating. If, according to our history, the wells had been built by the almighty rulers of antiquity, then decorating the lowest parts of the wells would be the most senseless thing to do because of their mostly submerged situation. These tire-like, ribbed stones were serving as coolants.

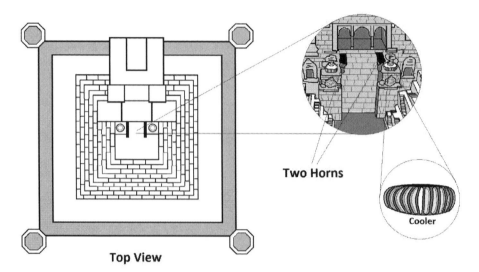

Two Horns

Cooler

Top View

Step Well of Chand Baori

(Pic. 61)

And again, as researchers we should not forget that we are not the first ones who discovered these marvelous structures. After the departure of the gods, many civilizations and hundreds of generations were using these stepwells as water reservoirs, and consequently every culture left its own heritage carved into the stones. Masterfully executed carvings on the walls of most ancient stepwells and temples are just additions to already existing structures of the post-satellite-ejection period.

Here we must admit that the imagination of the gods had no limitation. Probably experience in gold mining methods on different planets and the race for such a vital element as gold was pushing them to create such architectural diversity.

If we explore our planet further, we can see that the Great Masters were not limited by building square stepwells only. On the same Indian Peninsula we can find stepwells in the form of a ditch (Rani Ki Vav), octagon (Baoli Ghaus in Farrukhnagar, Haryana) and any other irregular shape (Nahargath fort step-well).

Gradually, by examining every structure of antiquity we can expand our observational area by leaving the Indian Peninsula and moving to the north-west. By reaching the regions surrounding the Mediterranean Sea we can observe the abundance of such ancient structures as amphitheaters. What we know about them is that they were built by our ancestors in an epoch of flourishing Greek and Roman cultures. But what we have never asked ourselves is: why would our ancestors have spent such great effort and resources to build these structures?

In many cases around these amphitheaters we cannot find any traces of other necessary structures for living. For example, at an elevation of 1,000 metres above present sea level, the Amphitheatre of Termessos in Turkey is located on the top of a mountain without any source of water. Any settlement there would be destined to dissolve due to its remoteness. If we mark locations of any large city or even village of our present or any other civilization, we can see that the majority of them are located on the shores of water. Whether it is a sea, lake, river or at least the foot of a canyon, it must have water to irrigate and cultivate any eatable crops. To suggest that every village of antiquity was marking its birth by building amphitheatres to gather everybody together does not seem very plausible at all. It seems like our ancestors were easily solving all the problems related to getting food, water and shelter and were now building centres of cultural communion. What nonsense.

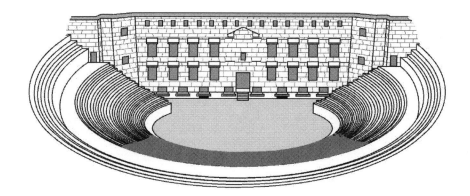

(Pic. 62)

All ancient amphitheatres, including the most famous Coliseum in Rome, were built to extract gold and gold only. One of the most compelling examples of that is the Amphitheatre of Aspendos in Turkey *(Pic. 62),* located 63 kilometres away from the Amphitheatre of Termessos where instead of a stage, our ancient architects decided to build a wall; a very strange feature for an entertainment facility.

In contrast to the Amphitheatre of Termessos, the city of Aspendos is located 30 metres above the present sea level. If we look closer at its amphitheatre we can see a striking similarity with the stepwell of Chand Baori in India. With many differences in shape and design, both structures served one purpose: to redirect water flowing from three directions towards the central wall for further filtration and extraction of tiny gold particles. An aerial view of this particular amphitheatre *(Pic. 63)* answers the question as to the purpose of the building and of all the amphitheatres and stepwells ever built on our planet.

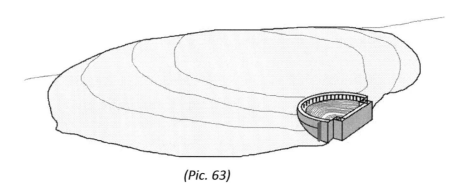

(Pic. 63)

Being located right on the edge of a small hill (which most likely has volcanic origins) it resembles a bitten off pie with a jaw stuck in it. The purpose of the amphitheatre built-into the body of the hill was the collection and further processing of water excreted from the fresh wounds of the Earth. That is the reason for such a large concentration of amphitheatres scattered around the Mediterranean Sea, which is the result of a massive crack in the Earth's crust.

Here we must admit that the amphitheatres scattered around the Mediterranean Sea have nothing to do with the extension of the Great Roman Empire. After the gradual healing of Earth's wounds, these gold extracting plants of the right-after-satellite-ejection period were later repurposed into the amphitheatres.

Returning to the Indian peninsula we can surely assert that the stepwells buried into the ground from all four sides misled us about their purpose. Their original role and location was to serve as a dam. Erected on the entire length of prehistoric river *(Pic. 60)*, these dams served as perfect gold extracting plants built to filter enormous volumes of water. Despite their similarity to our present day hydroelectric dams, the existence of ribbed coolers *(Pic. 61)* indicate passage of hot water, unless the ribs of those coolers are petrified blades of an electrical turbine.

If we look at the most famous amphitheatre of our civilization, the Coliseum in Rome, we can detect a mixture of ancient stones and considerably recent brick masonry. But what does not make sense in it is the existence of rows of trenches running along the entire arena. As in the case of the stepwells of the Indian Peninsula where all the structures of the right-after-satellite-ejection period were carved by the following generations of people, semi-round step-wells of the Mediterranean were transformed into amphitheatres with the following adoption of their architecture.

The same thing can be said about all the aqueducts scattered around the Mediterranean Sea. The structures of the antediluvian epoch adopted by the Romans became widely accepted by following generations. But let us scrutinize history. If the Roman Empire was so advanced in construction, then why was their main goal building amphitheatres and aqueducts? This is especially so on lands located far from Eternal Rome (which is present-day Spain, France, Turkey and Israel). "By conquering the lands of enemies, despite consumption of enormous resources, the Great Roman Empire was bringing their water supply technology to the people of Mediterranean." What nonsense.

During our history, no conqueror was ever involved in the construction of such grandiose facilities beyond the borders of his country. So far in our history, there were only two strategically important constructions on seized territories:

1. Defensive facilities to protect the conquerors' gains in case of rebellion.
2. Religious facilities to sustain their influence.

Neither the Great Roman Empire nor Napoleon, Hitler or any conqueror in our history had ever constructed any grandiose facility on seized territories because the main purpose of war is robbery and nothing else.

As in the case of the so-called Roman complex of Baalbek with its two layers of masonry completely different in size, our civilization inherited and reassigned the entire architecture of this gold extracting plant left by the gods. We cannot even imagine how much our architecture imitates the designs of our forefathers. Every religious architectural masterpiece of antiquity, whether it's a synagogue, church, mosque or cathedral — all of them are mimicking the Indian temples built by our creators. Similar to the powerful water jet in the centre of temples which were later transformed into deities, any religious structure on Earth has its own altar representing a pedestal of an item very important for that particular belief. Being unaware of our non-originality in architecture, we continue to build by the style left by our gods.

During the entire history, our great tendency towards art constantly camouflages originally built structures, and that is the main reason for our erroneous dating of artifacts left by previous generations. Every following generation adds something to structures already built by someone else, thereby proclaiming ownership of the entire structure as their own creation for generations to come.

If we traverse to the volcanically created Easter Island, then we can see that the same human tendency towards art had camouflaged the original purpose of the stone pillars scattered on the island. Here again, carved into them statues mislead us about their true purpose.

The tendency of numerous generations of aboriginal people who live on Easter Island to carve statues with the same facial features indicates just one thing. Before their transformation into statues, all the stone pillars on the island had the same geometrical characteristics eliminating the possibility of carving something else. As we can see, all the Moai statues have very distinctive sharp noses and chins, meanwhile the arms of these giants are disproportionately small and barely noticeable. Here we can surely assert that all these statues were carved from identically cut pillars, which could have been part of one facility.

Considering the island's volcanic origin, this area would not be missed by the gods who would definitely try to squeeze out as much gold as possible. The steady belching of geysers would be the perfect spot to gather the steam for further extraction of gold. Considering the island's remoteness from the rest of the world, it would be reasonable to build some kind of steam-capturing fence around the entire perimeter of the island. It is not coincidental that the majority of Moai statues are facing inland. According to their semi-buried and tilted condition we can definitely assert that Easter Island had gone through many landslides and therefore there could be many more statues, or maybe even raw pillars buried beneath the soil.

Now, knowing the processes of birth and death of every celestial body in the universe, we can make the following assertions: after reaching its peak, every planet eventually begins to lose its powerful shield against outer influences of the cosmos. With the gradually weakening shield, spectrums of ultraviolet and infrared lights that are dangerous for every living thing eventually penetrate the atmosphere. Death through the process of aging is inevitable.

During outer space exploration, our civilization constantly encounters problems with protecting its spacecrafts from the devastating influence of the infrared and ultraviolet lights of the Sun. Nowadays, the usage of gold to shield the crew and its equipment is the only effective method against intense infrared and ultraviolet lights. Of course, there are many other well-known IR and UV light reflectors (silver, copper, aluminum) but gold remains the champion among these elements because it is a metal that is non-corrosive and easy to handle. Considering gold's extraordinary ability to be stretched into one continuous sheet, it would be quite effective to wrap the planet in a golden foil to prevent the penetration of the deadly ultraviolet light. Of course, it would be very hard to cover the entire planet completely and leave it without any source of light, but launching a large fleet of satellites blanketing the planet would slow down the inevitable loss of atmosphere.

The gods' desperate need for a protective screen around a planet losing its atmosphere would drive them in search of gold. And when it comes to saving a planet, the scale of gold mining will also be of a planetary. Knowing the destiny of every planet, Earth was a perfect candidate in the search for relatively effortless mining.

Considering that the soil of the pre-satellite and even right-after-satellite-ejection periods of our planet consisted of more water-saturated compounds, all the prehistoric megalithic structures with enigmatically locked together masonry have a very simple and logical explanation. Before the ejection of satellites, the soil of Earth contained more water and therefore was much softer. It could take just a small effort to dig a couple of metres below the surface to reach a much softer tissue of the soil where, following exposure to the sun or heat, would lead to its gradual putrefaction.

Here we can make an assertion that all the megaliths were made from soft clay-type compounds that existed in abundance on Earth even during the immediately-after-ejection period. These structures were built on planet Earth completely different from our epoch, and our giant ancestors were using only locally extracted material. Common opinion that (connecting the stones) armatures were inserted into already carved spots is absolutely erroneous. The metal anchors were simply pressed into a viscous mixture.

Analogically with the monolith structures of South America, the monoliths of Baalbek were not quarried from the rock, they were dug out from a much softer clay-type soil where the following centuries of exposure to air had hardened them forever. The massive blocks of the great pyramids of Giza were not quarried and carried from hundreds of kilometres away; they were extracted right there from the Giza Plateau. Here it is necessary to mention that the similarity in construction techniques between South American megaliths and moulded casing stones of the Menkaure pyramid points at its exact date of birth. They were built right after the ejection of satellites where subterranean waters of the globe had not been drained towards lowland basins yet and the crust of the planet was much softer than it is now.

By analyzing these facts we can make the following assertions: if such massive structures were built on geysers, sinkholes or considerably small volcanoes, then we are talking about the right-after-satellite-ejection period when small-sized volcanoes were still hot enough inside. Even by assuming the possibility of survival of some civilized societies around the globe, earthlings would not have been able to consolidate themselves into such grandiose constructions. It would be more crucial for people to worry about their daily food rather than the extraction of quite useless (at that moment) gold.

We should not forget here that more than half of Earth's surface was ejected away and the newly-formed Earth was in the painful agony of its "postnatal" period. Much more powerful than today's earthquakes, geysers gushing around the globe and volcanoes would definitely degenerate any formerly-civilized survivors. This proves that all the megaliths of the post-satellite-ejection period were built under the leadership of the gods and for them only because no earthling would be interested in extracting gold on a planet devastated by hunger and thirst. Mourning their loved ones, earthlings were practically forced to urgently construct gold-extracting plants around the globe.

After such a disastrous event as the ejections of satellites, the only creatures capable of conducting such a grandiose construction would be the gods themselves who could survive the cataclysm by traversing to a nearby planet. Time was limited and the builders of these gold-eluting plants had to extract as much gold as possible because the power of the geysers gushing into the air would eventually diminish. This fact proves that all the prehistoric gold-eluting facilities around the globe belong to the same time period. In other words, the "mysterious" boulders scattered around the globe, dolmens, temples, amphitheatres, step-wells, pyramids and other prehistoric structures were built in the same right-after-satellite-ejection period and were built by the same race of gods to collect gold.

Offensive truth

As we go through many prehistoric structures of the Indian Peninsula, we cannot miss the famous erotic sculptures in the temples of Khajuraho, the Konark Sun Temple and others. Carved into the entire exterior of the temples' statues are silent remnants of a civilization, which had flourished here before the Great Deluge.

But what impresses us here the most is the protruding bulkiness of these statues. In other words, in order to achieve such a three-dimensional effect, our ancient masters had to hew out a substantial part of the walls. But would it not be easier and safer to install statues that were already carved instead of pecking them out from erected structures?

According to the unfinished condition of some of these sculptures *(Pic. 64)* we can make the assertion that all the carvings were carved out into pre-existing structures. Our civilization also uses similar methods of decoration, but unlike these three dimensional carvings our bas-reliefs of the Renaissance epoch were never carved so deep into entire body of the building due to the great risk of weakening its structural strength.

 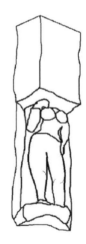

(Pic. 64)

In fact many carved temples of the Indian Peninsula do not look stable at all, where many extensively hewed out columns support quite massive ceilings. Excessive carving by ancient masons could have played a fatal role in the collapsing of one of the temples in Konark in 1837.

Besides fully carved complexes, the exteriors of many temples are extensively carved on their lower parts while the upper levels are left unfinished *(Pic.65)*.

Temples of Khajuraho

(Pic. 65)

And again, considering that all these carvings were made on existing structures, then all the parts of the walls untouched by humans retained their original look and therefore represent the architecture of the gods themselves. In other words, thanks to the fortunate decorative incompleteness of these walls, we modern people can still see and touch structures built by gods.

By examining the lower, extensively carved levels of the Khajuraho temples we can see that seemingly primitive erotic sculptures depicted on many Indian temples did not serve as some kind of sexual directory; they reflect the routine life of that period. But there is one, much more important factor there — the proportions. The sculpture carved on the exterior wall of one of the Khajuraho temples *(Pic. 66)* depicts a much larger male having relations with six much smaller but fully grown females.

Khajuraho, India

(Pic. 66)

A more scrupulous observation of this sculpture confirms Zecharia Sitchin's translations of ancient Sumerian cuneiforms. At first sight, the author's statement that "gods were taking daughters of earthlings as wives" sounds quite illusively romantic, but in fact the knowledge that the impregnation of the much smaller earthlings would eventually result in their death was not stopping these giants from their animalistic lust. The compatible genes of the two proportionally different races were causing mass deaths of earthy females during their pregnancy. The seed of the giant embryo would eventually overgrow the volume of intended space.

Now, Zecharia Sitchin's translations of ancient Sumerian tablets about gods who came to our planet in search of gold, their creation of humans as slaves and their later intercourse with earthly women sounds very probable. It is not coincidental that every language on our planet uses swear words directed towards the female. Nothing comes from nothing. I do not think that the present most offensive swearing on earth was a compliment during the reign of the gods, and we do not have any idea as to how old the "oldest profession" is and where it originated.

(It is a strange paradox that we praise our creators who used us as slaves. Our entire history was built on the belief in God as the highest level of perfection and purity, who created us "in his own image" and spirit. In the name of God, the Inquisition justified mass persecutions in Medieval Europe and the genocide of the entire Meso-American civilization; in the name of God numerous Crusades were launched; in the name of God people constantly take someone else's life today. The "blaming the third party" tactic accompanies us throughout our entire history and serves as a very effective political method to fulfill our selfish needs.

It reminds me of my childhood friendship. There were four of us back then, united and inseparable friends. When I met any of them separately, we were the closest friends. As soon as the third friend joined us, in a matter of minutes two of them were ganging up on the third one. It seems that the usage of a third party to counterforce another human being is imbedded in our genes.)

Of course, we can assume that this sculpture depicts the mating of individuals of the same race with different proportions but that is not the case. If we look closely, except the very obvious, this seemingly unremarkable sculpture has very important features.

- The male is twice the size of the females.
- The detailed depiction of the male's left palm — his hand has six fingers.
- Along with such a detailed depiction of the belly buttons of all the females, the male has no sign of a belly button whatsoever. If the reader assumes that the male in this scene wears clothes, then I would have to disappoint you. The original sculpture has all the details of a human body (deliberately censored by me) including exposed genitals.

If the sculptor of this explicitly detailed carving did not miss the lines on the male's palm then he could not miss one of the main features on our body – the belly button. This barely noticeable detail gives us an unimaginable insight into the origin of the gods themselves. Unlike humans, gods who came to Earth from a different planet could have a reptilian origin. What comes to mind here is God's curse after Eve's disobedience:

"I will greatly multiply thy pain and thy travail; in pain thou shalt bring forth children..." (Genesis 3:16)

The fact that (after some action unknown to us) Eve and then Adam were transformed into mammals points at their non-mammal origin. God's curse of giving birth in pain means that it was not planned originally and the consumption of an unknown matter from the *"Tree of the knowledge of good and evil" (Genesis 2:3)* triggered the drastic mutation of probable reptilians into mammals. It is hard to assert but reptilians that lay eggs should experience less pain than viviparous mammals that give birth. After the miraculous transformation:

"And the eyes of them both were opened, and they knew that they were naked; and they sewed fig-leaves together, and made themselves girdles." (Genesis 3:7)

Now, let us suppose that before their mysterious transformation, Adam and Eve had no visual differences on their bodies, just like reptilians whose sex organs are hidden inside the body. The absence of differences means the absence of antonymous concepts all together. In other words, we wear clothes to cover the differences between the male and female parts of our bodies. Men can walk on the streets with naked torsos because independent of their sex, all the children have flat chests and our eye gets used to seeing that. The development of breasts in teenage girls leads to the necessity to cover them in order to hide the differences between the two opposite sexes. Identically we tend to hide all the defects of our bodies in order not to stand out from the rest. So the fact that Adam and Eve were not ashamed of being naked is indicative of their visually identical bodies. The sudden development of mammal features surely led to their physical differences and the appearance of such feelings as shame.

Now, a little bit of science fiction. In our example, we know about the gods' accomplishments in genetic engineering, but we have no idea about their space travelling technologies. If travelling from planet to planet takes more than a lifespan, it would be very rational to create humans who would be able to lay eggs instead of being viviparous. Sending mammal-like humans into space carries many difficulties in providing the maintenance necessary for the crew. To dispatch hundreds of ready-to-hatch eggs would allow controlled birth of every new member of the crew.

The sudden departure of the creators can be explained by a drastic worsening of their planet's atmospheric (and thereafter political) conditions. We can only imagine the situation on a planet gradually losing its atmosphere. Like the sinking Titanic, it would provoke fear, hatred, betrayal and chaos among its inhabitants. The gods' promise to return soon was based on the hope of evacuating their loved ones to Earth. Unfortunately, we know nothing about their fate. Did they save their planet? Did they stay here to observe and protect us or did they flee to the nearest galaxies? We can only guess. One thing is certain, after their departure, the structures they erected were not destroyed and were continuously producing gold-rich mixtures for centuries to come. That is the reason of such lavishness of golden artifacts in the ancient civilizations of India, Central America, Egypt and China.

- Such an easy and quick filling of an entire room with gold in exchange for their imprisoned king, Atahualpa, was achieved by the Incas thanks to the continuing process of gold elution after the gods' departure.
- No ancient emperor of Eurasia had the capability to cast a golden sarcophagus for himself other than Egyptian pharaohs.
- No religious facilities store so much gold in them as the temples of India.

The incomprehensible fascination with gold was mysteriously transmitted to earthlings from the gods themselves. Considering that the gods were using humans as the main mining labour, every gold nugget ever found by an earthling would be generously rewarded. Would it be in the form of food, temporary release from work, or even permission to be with a woman; all these privileges would definitely encourage earthlings to work harder. That period can be marked as the beginning of the bloody history of gold among humans because the discoverer of any substantial nugget would risk being killed by others in order to claim the desirable reward.

The same period can be considered as the cradle of our monetary system. With no doubt, people began to exchange gold among themselves, which generated furiousness among the gods and their cruel punishments of earthlings. The entire human history is built on the reward and punishment principle. Nothing comes from nothing. The very foundation of human society was embedded by the gods themselves. That is the reason for the gods' gradual disappointment with their beloved creations and that is the reason for the gods' relentless desire to destroy mankind. But duty and the willingness to save their own planet was holding them back from this relentless act.

Atlantis

All the major footprints of the former poles on Earth were created by severe Ice Ages during the moonless existence of Earth. We know that the existence of satellites around the planet and alternation of seasons softens the Ice Age effect by many times. With no access to oceans, snow accumulating on dry land continued to build up its "icy kingdom" by absorbing the entire vapour from the air. Similar to the formation of the "Vampire Ocean," the precipitation and accumulation of snow on the poles located on dry land have led to the eventual drainage of near equatorial waters. The only condition for Earth to undergo such a major ice age would be the situation of one of the poles on one of the continents.

After the First Great Catastrophe when our planet lost a substantial part of its crust and acquired a satellite stabilizing its own rotation, our globe established its considerably permanent location of poles. In other words, due to the "bitten off" shape of our planet, the geographical poles of our globe are destined to be in the vicinity of the present Arctic and Antarctic regions.

There are two major facts proving the plausibility of this statement:

> 1. Frozen mammoth remains in Siberia indicate their constant closeness to the geographical pole.
> 2. Our civilization's two main written records of the Great Deluge that have survived (*The Chronicle of Akakor* and the Old Testament) are located in the Andean Mountains and the

Middle East accordingly. This is a direct indication that after the First Great Catastrophe, Earth's geographical poles were never located in those two regions.

Knowing the past of our planet, we can link it to the descriptions of our ancestors, who used to live in that seemingly eternal antediluvian epoch.

According to Plato, Solon had heard the story of Atlantis from an Egyptian priest who stated:

The [Atlantic] ocean there was at that time navigable; for in front of the mouth which you Greeks call, as you say, "the Pillars of Heracles" [Hercules], there lay an island which was larger than Libya and Asia [Asia Minor] together... Yonder is a real ocean, and the land surrounding it may most rightly be called, in the fullest and truest sense, a continent. Now in this island of Atlantis there existed a confederation of kings, a great and marvelous power, which held sway over all the island, and over many other islands also and parts of the continent; and, moreover, of the lands here within the Straits they ruled over Libya as far as Egypt, and over Europe as far as Tuscany." (I. Velikovsky "Worlds in Collision", page 154)

Here again, we have a large discrepancy between ancient legends and our reality, and our first reaction is blaming Plato for making this story up. But all we have to do is link his description with the reality of that time.

When we read somebody's description of "something being located beyond some point," after reaching it, we always expect to find it in front of us. For example, a broom located behind the door is not necessarily right behind the door. It can be located on the right or on the left of it. The main idea is that its true location is behind the door. That is the reason of our misunderstanding of Plato's story because the meaning of beyond *"the Pillars of Heracles"* clearly suggests that the legendary island is located beyond the Mediterranean Sea.

Knowing the antediluvian landscape of the world, we can easily unscramble Plato's story. First of all, the original story comes from the *Egyptian priest* who obviously lived in Egypt. If we eliminate the possibility of air travel in antiquity, then sea travel would be the only way to reach any island. Here it is quite logical to read the priest's description of the legendary island located beyond *the Pillars of Heracles* (present-day Gibraltar) because any ancient Egyptian resident who had ever travelled beyond the Mediterranean Sea would undoubtedly pass present-day Gibraltar.

A passenger of any marine vessel would describe the location of the famous (at the time) "Island State" as "beyond Gibraltar" and a non-educated storyteller would never describe the destination after passing the strait. But people's logic is quite rigid, and very few of today's sea cruise passengers would describe their port-to-port trip as "first we sailed north-west and then south." The majority of us would just list the seaports we have visited during our voyage. For example, for a sea traveler whose itinerary is Egypt to Britain, the description of Britain's whereabouts would be "beyond Gibraltar."

The next step in unraveling the past is to determine the location of the poles and our biggest help are ancient maps. One of the best helpers in our task is the so-called mysterious map of French mathematician and cartographer Orontius Finaeus (1494 – 1555). There are two main puzzles in his maps: the precise depiction of present continents and the strangely different from the present shape of polar ice caps *(Pic.67)*.

As we can see on the comparison maps, the recurrent shift of poles is not just the relocation of geographic poles. Due to the wide diversity of landscapes, poles periodically travel around the globe creating a unique distribution of snow, fundamentally changing the face of the earth. That is the reason for such substantial differences between the map of Orontius Finaeus and the face of Earth at present.

The first enigma here is that back in the 15th Century our civilization's cartography was not as advanced as nowadays and the author of the map could not have possibly known the outlines of all the continents, islands and oceans in such a precise manner. Even if we were to suggest that the author of the map had the ability to travel around the globe on some kind of flying apparatus, his map has one very distinctive feature: the size of Antarctic ice. Such an impressive size of the Antarctic continent can indicate one thing only: the Earth depicted on the ancient map was experiencing a major Ice Age. Considering the enormous size of the

Antarctic continent, we can make the conclusion that Orontius Finaeus redrew his map from a much earlier map that is unknown to us.

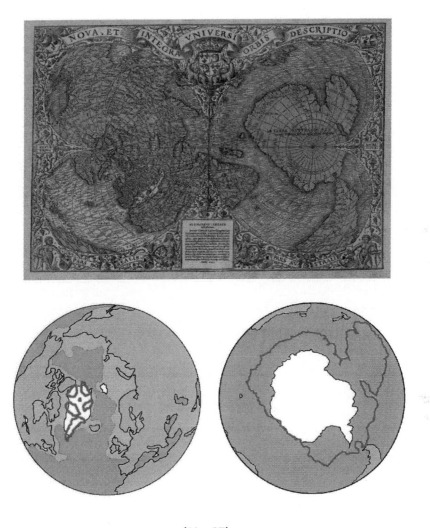

(Pic. 67)

If we look at the map closely, we can conclude that in order for the Antarctic ice to accumulate such an immense mass of snow capable of closing the gap between the southernmost tip of South America and the Antarctic continent, the pole has to be located somewhere between these two continents. And if we mark the spot between these two continents as one of the poles, then the second pole appears somewhere at the Bering Strait. The author of the map showed no sign of any strait whatsoever, joining the two continents into one united mainland.

Such an unfortunate scenario for every living thing on Earth brought it to a complete freezing of the shallow Bering Strait and a massive ice bridge between the southern tip of the South American and Antarctic continents. The concentration of such a large quantity of snow on the poles means severe drought on the remaining parts of the globe. Here we should not confuse this period with the medieval "Little Ice Age" because Europe at that time was experiencing substantial cooling, and the absence of any water mass on the map between the Eurasian and North American continents indicates the nearness of a geographic pole to the present Bering Strait, which is much farther from Europe than the present pole.

A more detailed comparison of the present world with the other antediluvian map of Orontius Finaeus reveals obvious similarities *(Pic. 68)*. At first sight, the map has many discrepancies, but overlapping the modern map of iceless Greenland with the map of Orontius Finaeus reveals a startling discovery. During the Ice Age, regions distant from the poles had experienced severe drought, and if we virtually eliminate Greenland's ice and pump all the near coastal waters away, the outlines of the newly shaped continent will coincide perfectly with the shape of the islands on the map.

(Pic. 68)

As mentioned by Plato's priest, the island of Atlantis that was "*larger than Libya and Asia*" is undoubtedly still the largest island of Greenland.

The legendary *"confederation of kings"* could possibly rule over vast territories of the present Canadian Arctic Archipelago, Europe and North Africa. One of the proofs that this map does not belong to the epoch of Orontius Finaeus and depicts the period of the antediluvian Ice Age lays in the completely wrong designation of all the islands *(Pic. 69)*.

For example, drawn by Orontius Finaeus island of Novaya Zemlya has completely different and rounded outlines. A close observation of this map reveals numerous discrepancies of designated lands and islands. For example, little could he know that the mountains redrawn by him on the Eurasian continent are indeed highlands of the present island of Novaya Zemlya. Considering that the antediluvian ocean level was much lower than now, many islands presently close to the continent were once part of it. The comparison map below shows lands realistically corresponding to each other.

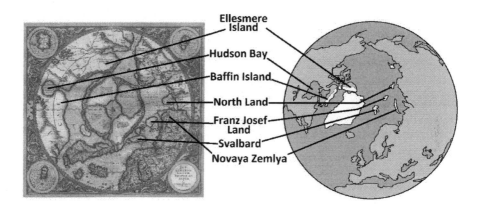

(Pic. 69)

The gradual receding of seawaters of the planet experiencing a major Ice Age was opening vast areas of land promoting the farther spreading of humans around the globe. Discoveries of submerged structures close to the present seashores indicate their belonging to the antediluvian civilization that lived and thrived on the coastal lines completely different from the present.

Such precious discoveries of submerged megalithic structures near the coasts of Cuba, Japan and the Mediterranean region indicate their

185

different timing in the chart of human existence. Before, when we were reading Plato's story about the majestic island-city surrounded by a moat, we were imagining some extraordinary ditch dug around the city and filled with water as protection against foreign invasion.

Here again, we have the wrong perception of our far ancestors. The entire globe was scattered with volcano-type islands fallen to Earth from the ejection of satellites during the First Great Catastrophe. After massive eruptions of lava, the tops of many volcanoes just collapse into the cavity beneath, creating perfectly round rings of water. If Mother Nature had sculpted such a perfect shelter, then there was absolutely no reason for humans to spend such enormous effort to construct it by themselves.

Now, knowing the period when the original map was drawn, we can surely assert that Orontius Finaeus was not the author of this map, and was an excellent cartographer who redrew it from an antediluvian map that miraculously survived. Thanks to Orontius Finaeus and other cartographers of the past, we can fully reconstruct the order of events preceding the Great Deluge.

According to Orontius Finaeus' map we can definitely assert that the Arctic Pole that was mistakenly marked by the author (the small island surrounded by four large islands) is in fact the most mysterious and legendary island of Atlantis. Knowing the real events of antiquity, all the mysterious cities do not seem so mythical any more. If, during the following recurrent reversals, our Earth ever enters an Ice Age, the massive evaporation of the oceans will gradually reveal the sunken cities of antiquity along with the legendary city of Atlantis.

One thing that is absolutely certain is that the legendary island of Atlantis surrounded by a ditch is located in Greenland, and the only possible way to rediscover its remains is to wait for Greenland's complete release from the snow. Who knows, maybe a more moderate regional climate will be dictated by our planet's next recurrent reversal.

Recent Polar Shift

The disappearance of Atlantis and many other civilizations of antiquity is not a singular and finished process. Our planet periodically experiences polar shifts and, as we mentioned above, the pole was, is and will always be in the vicinity of the Arctic Ocean. Whether it be in the middle of it, in Greenland or close to the Bering Strait, it will always be around. And if we think that its cyclic recurrence takes thousands of years, then we are deeply wrong. One of the best examples is Europe's Little Ice Age, which, according to historical data, lasted from the 16th to 19th Centuries.

Of course, it is hard to define the exact time range of that period, but thanks to many great constructions of the 18th Century, we can create a more realistic picture of that epoch. One such grandiose construction is the Ladoga Canal built by Peter the Great, Tsar of Russia from 1682 until 1725. Having the total length of 117 kilometres this marvelous structure surprisingly duplicates the coastal line of Ladoga Lake *(Pic. 70)*.

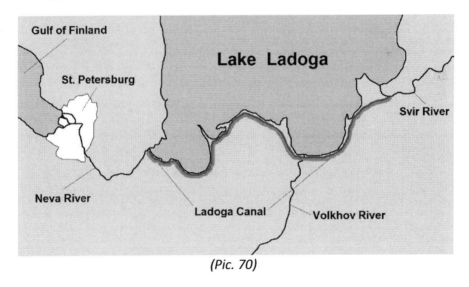

(Pic. 70)

Knowing the imperial ambitions of Peter the Great, who devoted his life to the expansion and strengthening of the Russian Empire, building the "Window to Europe" city of St. Petersburg was his priority. Apparently, building the Ladoga Canal was dictated by the necessity to do so and was not just his monarchic caprice. Modern sources suggest that the Ladoga

Canal was built for the safer sailing of cargo ships. According to that statement we must build similar canals on the perimeters of all the present oceans. What nonsense!

Here again, we think that our ancestors had a lesser level of intellect than present people. The decision of Peter the Great to build the canal in 1719 was dictated by the necessity to get a stable waterway to connect the newly-built city on the Neva River with the rest of the kingdom. By connecting the waters of Svir River with the apparently diminishing waters of Neva River, the city of St. Petersburg would be provided with a stable waterway connection to the rest of Russia.

If we look at the map above, we can see that the Neva River presently outflows from Lake Ladoga. If "His Majesty" decided to connect the Svir River (that flows into the Ladoga Lake) with the Neva River (that outflows from the Ladoga Lake), that means the Ladoga Lake as we see it today was not in existence. In other words, the lake's shoreline was further away from these rivers presently linked by the lake.

Considering the above, we can make an assertion that at the time of construction of the "Ladoga Canal" in 1719, the North Pole was either located on dry land or had recently shifted to its present location because only that condition could lead to the drainage of many lakes of the planet.

According to the frozen condition of the island of Greenland at present, we can surely assert that the previous location of our present North Pole was Greenland. As we see, the poles of our globe could have shifted to their present location just 300 to 500 years ago, and therefore the topography of our planet at that time was completely different.

The Second Great Catastrophe

The Drought

In his book *The 12th Planet*, Zecharia Sitchin mentions a severe drought preceding the Great Deluge.

From above, the heat was not... Below, the waters did not rise from their sources. The womb of the earth did not bear; Vegetation did not sprout... The black fields turned white; The broad plain was choked with salt. The resulting famine caused havoc among the people. Conditions got worse as time went on. The Mesopotamian texts speak of six increasingly devastating sha-at-tam's — a term that some translate as "years," but which literally means "passings," and, as the Assyrian version makes clear, "a year of Anu": For one sha-at-tam they ate the earth's grass. For the second sha-at-tam they suffered the vengeance. The third sha-at-tam came; their features were altered by hunger, their faces were encrusted... They were living on the verge of death. When the fourth sha-at-tam arrived, their faces appeared green: they walked hunched in the streets: their broad [shoulders?] became narrow.

By the fifth "passing," human life began to deteriorate. Mothers barred their doors to their own starving daughters. Daughters spied on their mothers to see whether they had hidden any food. By the sixth "passing," cannibalism was rampant. When the sixth sha-at-tam arrived, they prepared the daughter for a meal: the child they prepared for food... one house devoured the other. (Page 392)

The devastation of people caused by drought is hard to comprehend. Here, Zecharia Sitchin insists that the meaning of "sha-at-tam" is the year of Anu, which is quite logical for visitors from a different planet to continue to use their own timing, which is the calendar of their mother planet. Unfortunately, we do not yet know the duration of a year on planet Nibiru to compare with our own. What we can assert for sure is that one sha-at-tam cannot be equal to 3,600 Earth years.

The Rain

"And the rain was upon the Earth forty days and forty nights."
(Genesis)

For us modern people, the existence of such continuous rain seems more mythical than realistic, but let us scrutinize it again. We know that in order to get an Ice Age effect, at least one of the poles has to be located on dry land. Prior to the flood, Earth experienced a severe Ice Age and drought. During the Second Great Catastrophe, the sudden shift of Ice-Aged poles towards the equator led to their intensive evaporation and following torrential rains around the globe accordingly.

As we know, during the First Great Catastrophe (the ejection of satellites) all the flora and fauna whose habitat was in the vicinity of the present Arctic regions of our globe were swept away by enormous waves and were frozen forever due to their sudden dislocation towards the pole. The existence of perfectly preserved animal remains on the present Arctic north from the times of the First Great Catastrophe indicates their constant closeness to the geographic pole. The suggestion that they belong to the time of the Second Great Catastrophe is very doubtful because there have been no mass graveyards of mammoths or any other prehistoric animals ever found in Hudson Bay Basin and the Canadian Archipelago. That fact indicates that the 2000 - 3000 kilometre diameter ice cap of the pre-satellite-ejection epoch had prevented the penetration and deposition of any organic matter. So, what we can assert here is that the antediluvian pole was located somewhere close to present-day Siberia, and its rapid melt down could be possible only in one case — the recurrent reversal of Earth with the following shift of its poles to the equator.

The fact that the majority of animal corpses found in Siberia are still in a frozen state indicates that during the Second Great Catastrophe, the poles that were rapidly dislocated towards the equator did not stay there for a long period of time, and after extensive evaporation of ice located on the equator, the balance of the globe reached its critical point, and the Earth returned to its original (or close to original) position because after the ejection of satellites, the "bitten off" shape of the globe acquired a considerably stable centre of gravity.

In his book *The 12th Planet*, Zecharia Sitchin describes the gods' retreat to a near-Earth orbit right before the arrival of the Deluge. Just the fact of their presence on Earth in that particular epoch indicates the closeness of their planet Nibiru and our "second star" accordingly. We do not know whether the appearance of our second star provoked any spark between the stars, but the close passage of such a massive star could easily cause a mass displacement of geographic poles of other planets.

Whether it was the result of a recurrent reversal or the passage of Nibiru, in both cases the sudden dislocation of polar ice accumulated during the Ice Age towards the equator would lead to its accelerated evaporation and massive rainfalls accordingly. If the legendary Noah's Ark landed on the mountains of Ararat, then despite its definition as the Great Deluge, the flood had both a worldwide and local character at the same time. It was not the sudden breakthrough of a high-altitude ocean like the "Vampire Ocean" during the first catastrophe. It was exactly as it was described in the Old Testament: massive continuous rainfalls for forty days and forty nights, which led to the accelerated filling of all the basins of the Earth.

Prior to the massive rainfall, Earth was experiencing severe drought. Animals and people who were following constantly shrinking lakes and rivers were living in the lowlands close to constantly receding waters. It looks like Earth was purposely entrapping all the living things together to destroy them all at once. That is the main reason for such a devastating impact on Earth's human population, as well as on flora and fauna.

Entire civilizations were swept away by constantly arriving waters, landslides, earthquakes and awakened volcanoes. The intensive melt down of ice and erosion of soil from the face of the Earth stripped all the sleeping volcanoes buried under the weight of the soil. That is the reason for the spontaneous awakening of all the volcanoes of the planet.

Precious information from the Old Testament about *"forty days and forty nights of rain"* gives us an important clue to events which took place in antiquity. Just the possibility of the existence of such intense and continuous rainfall indicates massive melting and rapid evaporation of ice accumulated during the preceding centuries of a continuing Ice Age. The number of continuously raining days, forty days and forty nights, indicates the real scale of that particular catastrophe and is not related to superstitions associated with number 40.

The majority of civilizations vanished in this Great Deluge, but the fact that there is no mention of Atlantis in the Old Testament indicates that Noah and his family were not the only ones who survived this cataclysm because the story of Atlantis had survived through different sources of antiquity. We cannot tell whether Noah ever heard of Atlantis, but we can certainly assert that Noah and the people who lived and drowned in the city of Atlantis had lived in the same period of time, that same year.

The Ark

As commanded by Enki, -Atra-Hasis sent everybody aboard the ark while he himself stayed outside to await the signal for boarding the vessel and sealing it off. Providing a "human-interest" detail, the ancient text tells us that Atra-Hasis, though ordered to stay outside the vessel, "was in and out; he could not sit, could not crouch... his heart was broken; he was vomiting gall." But then:...the Moon disappeared...The appearance of the weather changed; The rains roared in the clouds... The winds became savage... the Deluge set out, Its might came upon the people like a battle; One person did not see another, They were not recognizable in the destruction. The Deluge bellowed like a bull; The winds whinnied like a wild ass. The darkness was dense; The Sun could not be seen. ("The 12th Planet" Page 403)

According to the Old Testament, Noah was the chosen one to complete the mission. But knowing the scenario of the upcoming disaster, I would suggest that the selection of Noah was based on the geographical location he was living on and his skills. The probability of survival on the near ocean lowlands would be close to zero. Survival in the mountains would bear risks of being caught in a mud flow and landslides. Besides, would you trust the building of such an extraordinarily large ship to a resident of a mountainous region who has never even seen any sea? The most suitable place to survive would be near some highland water basin with an abundance of building material — forests. And of course, the builder of the ark should have been familiar with life at sea and vessel building techniques. What assertion we can make here is that besides the considerably safe habitat, Noah was chosen for his carpentry and naval skills.

The global drought preceding the catastrophe forced all the coastal settlements to migrate towards receding waters and then, one day, the long-awaited rain came down to Earth. At first, people starved from continuous drought would react with joy and happiness at the long-awaited end of their torments and the beginning of florescence. But vain hopes soon began to crumble because the stormy rain was not ceasing. Retreat to the previous dwellings located away from an already furious lake could not save anyone because they were also destroyed by the torrential bombardment. The absence of caves or any other shelters in the landscape surrounding the lake would sentence every living thing to death. The possibility of survival in such a sunless environment for weeks to come is very doubtful.

"Fifteen cubits upward did the waters prevail; and the mountains were covered." (Genesis)

Considering that the ancient unit of measure — the cubit — is the length from the elbow to the tip of the middle finger, we can assert that waters of the Deluge rose seven metres. From the modern viewer's point of view seven metres does not seem so catastrophically critical, but in order to feel that horrible event, we have to traverse into that epoch.

Here we must mention one quite important aspect. Of course, all the deep-water measurements on board the ark were done either by crew members or by Noah himself. The only way to measure the depth of the water overboard is to use weight tied to a rope. And of course, in order to measure the length, any human being would use an almost universal measuring tool, which is always handy — a part of his body. (As a person who grew up in the metric system of measurements, I remember, in childhood we were using our feet in order to mark equal widths of soccer goals). Modern people are still using feet as an official unit of measurement. Therefore, it sounds very logical that the ark's crew was measuring the length in cubits. This ancient unit of measurement has survived until our days with one exceptional difference: antediluvian people were much taller than we are today, which means that 15 cubits for Noah may possibly be equal to the present humans' 20, 25 or even more cubits. So, the waters of the Deluge could possibly rise as much as 10, 13 and even more metres.

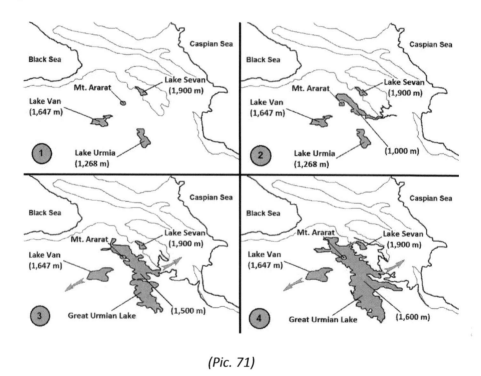

(Pic. 71)

But according to the flood imitation map *(Pic. 71)*, in order for the newly appeared lake to cover the base of Mt. Ararat it could not just gain

194

an extra 10 or 20 metres of water. It is supposed to be elevated by hundreds of metres. Here we can have two answers:

- Either there were no more available ropes on board the ark and the following measurements were impossible to conduct

- Or the present equivalent for an ancient cubit is erroneous

If, according to the Old Testament, Noah's Ark landed on the mountains of Ararat, then it was lifted by flooding waters from the basins close to Mt. Ararat. That gives us precious information about the birthplace and manhood of Noah himself. During the severe drought preceding the flood, people had to follow receding waters of rapidly evaporating lakes. At present, there are just three plausible candidates in the vicinity of Mt. Ararat to be considered as Noah's homeland: the basins of Lake Van, Lake Urmia and Lake Sevan.

A virtual simulation of a severe flood in the vicinity of Mt. Ararat clearly indicates that expansion of lakes Van and Sevan could never reach the biblical mountain; therefore, they are off the list. Lake Urmia on the other hand has all the chances to surround Mt. Ararat due to its location in the same basin. Considering that present Lake Urmia is saline, we can assert that it existed way before the Deluge and even before the First Great Catastrophe. Along with Lakes Van and Sevan, Lake Urmia had been scooped up by the elevating equatorial bulge. In contrast to Lakes Van and Urmia which have no water outflow, Lake Sevan has a river draining into it — the major player in converting salty lakes into fresh water.

A closer study of the area indicates that the landscape surrounding Mt. Ararat is just one large basin capable of containing a large volume of water. The map below *(Pic. 72)* depicts a virtual flood of the Ararat Plain. Of course, along with filling up the highland basins of the region, the sea level of the entire planet was drastically raised. For example, increasing the water level by just 100 metres would join the present Caspian and Black Seas together, but we do not want to consider other flooded regions in order to concentrate on the biblical event.

The continuous and extensive rain would expand the newly appeared lake farther and farther, filling the entire highland basin with water. Due to its formation on the same water basin as Lake Urmia, let us name it the "Great Urmian Lake." But there is only one possibility for this lake to reach such high altitudes: the basin should be completely enclosed.

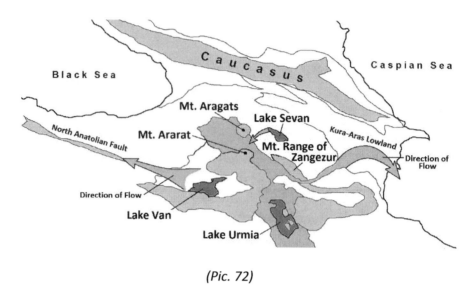

(Pic. 72)

According to the present landscape of the region, the floodwaters would escape through the riverbed of the present-day Araks River towards the east, and the lake's water level would never exceed the 1,000-metre mark. In order for waters of the "Great Urmian Lake" to reach their maximum (which is 1,600 metres above the present-day sea level, at a higher altitude the lake would have more than one out-flowing channel), the present Mountain Range of Zangezur would not be divided by the riverbed of Araks River. In other words, the antediluvian landscape of the region should be just one large basin of the "Great Urmian Lake" without any out-flowing rivers. Proof of that lays in the present chemical composition of soil in the Ararat Plain, which is described as saline-alkaline.

Being under enormous pressure from the overflowing lake, the part of the Zangezur Mountain Range collapsed, which led to the sudden and impetuous escape of the lake's water towards the Caspian Sea. Such spontaneous drainage of the basin via the present Kura-Aras lowland made an enormous curved indent on its landscape, a footprint of streaming down water. Carrying miscellaneous debris, animal and human corpses, the waters of the lake eventually reached the Caspian Sea depositing all its deadly cargo on the bottom.

Along with the overfilled basin of the "Great Urmian Lake," overfilling of the other two lakes would also lead to their massive drainage. The water excess of Lake Sevan was adding to the already overfilled basin of the "Great Urmian Lake" and waters of the present-day Lake Van were flowing down via the "North Anatolian Fault," carrying all the dirt and debris towards the Black Sea and depositing them in the end of the "North Anatolian Fault" creating quite smooth lowland.

As we know, the perfectly smooth surface of any landscape can be achieved by the sedimentary nature of a water body only. And the Ararat Plain is definitely one of them. If we eliminate the possibility of the Ararat Plain being submerged under water before the Great Flood, then its plain landscape could be formed by the waters of the Great Flood itself. In any case, at some point in time, the Ararat Plain was submerged under the saline waters of the "Great Urmian Lake."

According to the flooding map of the area we can try to determine the birthplace of the legendary ark.

- At an altitude of 1,750 metres Lake Van has out-flowing drainage directed away from the newly appeared lake. This fact eliminates the possibility of any vessel crossing the highlands separating the two lakes.
- If we are to suggest that the Ark was built on the shores of present-day Lake Urmia located 170 kilometres away from Mt. Ararat, then the out-flowing drainage of the lake located between them (the gap in Zangezur Mountains) would carry the vessel downstream towards the Caspian Sea rather than closer to Mt. Ararat. Of course, it is hard to predict the directions of each stream in the lake, but the possibility of the Ark's travel to Mt. Ararat alongside a powerful downstream is very unlikely.
- Located north-east from Mt. Ararat, Lake Sevan could be a probable candidate in the list. If we suggest the Ark's birthplace be the shores of the antediluvian Lake Sevan, then it would be quite a harsh journey for its passengers on its way down towards the mountains of Ararat. This fact does not decisively eliminate this possibility but travelling downstream in an uncontrolled vessel would definitely bring it to its disintegration.

Recent archaeological discoveries of a prehistoric metallurgical foundry in Metsamor, Armenia revealed quite an impressive complex of smelting furnaces capable of processing iron, copper, bronze, zinc, mercury, manganese, strychnine and gold. Located right on the border of the Ararat Plain and therefore being the most accessible for excavations, the archaeological site of Metsamor is just the tip of the iceberg of forever vanished cities buried beneath the entire Ararat Plain. Taking into account the fact of the global drought preceding the Great Deluge, the main population of the antediluvian civilization in that region would be concentrated closer to the source of water, which would most probably run in the middle of the present Ararat Plain. We can grievously imagine the settlements and villages of antediluvian civilizations buried beneath the plain — ancestors of Noah and our own accordingly.

In 1943, during the foundation construction of the "Victory" bridge in Yerevan (the capital of the Soviet Republic of Armenia back then), builders came across a metal pillar of unknown origin sticking out from the ground. Extraordinary qualities of the pillar did not allow builders to break or even chip a piece from it for further examination. After unsuccessful attempts to withdraw the pillar from the ground and the urgency of the bridge construction during the war, the artifact was buried into the foundation of the future bridge forever. Having a description of the artifact's qualities similar to the famous pillar of India makes us conclude that both pillars were smelted in the antediluvian period, where the oxygen level was much higher than at present. Being located just 32 kilometres away, the pillar buried in the present-day bridge's foundation was most probably smelted in the furnaces of the present Metsamor archaeological site.

According to the above mentioned archaeological findings, we can certainly assert that right before the Great Flood, due to global drought, the area of the present Ararat Plain was completely or partially dry. But being the largest water basin in the region during the drought it would have more chances to sustain precious water in it than other lakes. In order to prove or disprove any possibility we need to find the final resting place of the Ark.

In search of the legendary Ark we must mention the name of Ron Wyatt (1933-1999), the American adventurer, who was devoted to proving the authenticity of Noah's Ark remains at the Durupinar site in Turkey *(Pic. 73)*.

The first impression we get when we look at this site is that it is a natural formation, but for some reason it is located right between two hills, where an immense mass of water receded through this passage and could quite possibly carry the ship with it. It is located at 1,968 metres above the present sea level. Slopes pointed upward on the perimeter of that formation can be explained not as remains of the ship itself, but as fossilized earth deposited around the boards of the ship by receding waters.

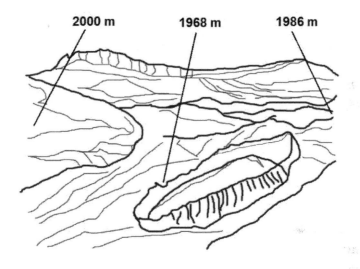

(Pic. 73) Suggested resting place of Noah's Ark near Uzengili, Turkey

If we are to suggest the authenticity of this site as the final resting place of Noah's Ark then we can declare the following facts:

1. The site is located on the east side of the highland facing Mt. Ararat from its west, and the first elevated landscape to be seen from there would be the peak of Mt. Ararat.
2. To try to find the remains of the Ark would be in vain as landing on the slopes of the wiped off highland would require the gradual demolition of the vessel by its survivors in order to use the wood as building and burning material. We should not forget that the receding waters left only mud deserts.

Whether this site was the real landing place of the ark or not is not so important. The most important feature here is the landscape that surrounds it. This unique formation and area around it was sculpted by receding water. The altitude of the Durupinar site is 1,968 metres above the present sea level, and such a massive mudslide could be caused either by an enormous wave or by the biblical torrential rain. We can certainly accept that this formation was created by a massive mudslide, but if we look at the area from above, there is another quite interesting feature *(Pic. 74)*.

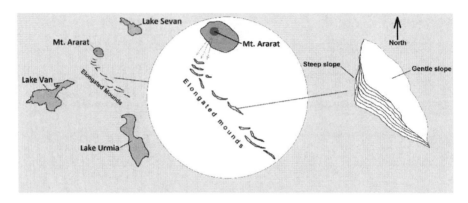

(Pic.74) Rising from the surrounding landscape, elongated hills have an average height of 200 metres.

A satellite view of the area shows some orderly directed hills located on the south-west of Mt. Ararat. Outlining the south-west borders of the former "Great Urmian Lake," three of these parallel hills have a curved shape, and the virtual centre of these hills points directly at Mt. Ararat. All the hills have one similarity: a gentle slope on the east and a steep slope on the west — an obvious indication of a massive wave from east to west. The height of these hills varies from 150 to 250 metres above the surrounding landscape. Here we can try to build a logical chain of thought:

- If these hills were created by a volcano of Ararat awakened in the water, then all of them would be directed away from its epicentre, but they are not.
- Of course, it is hard to comprehend the power of an erupted volcano, but in order to lift, carry and deposit 200-metre-high mounds, the wave itself should be taller many times over.

There are only two possibilities for these hills to be formed, and in both cases those are footprints of enormous waves:

- By the waves of impacted debris from ejected satellites.
- By a wave of the sudden shift of the planet to its original position at the end of the Great Deluge because that was the only time when the basin of the "Great Urmian Lake" was filled to its maximum. As we know, rapid evaporation of the ice of the former poles and their massive showering on the rest of the planet led to a rapid change in Earth's balance, and the globe shifted back to its optimally stable position. And again, such a rapid shift would cause the occurrence of massive tsunamis in overfilled basins.

Despite the obscurity of the exact cause of these formations we can make an assertion that these hills were formed by waters of the "Great Urmian Lake," the motherland of Noah and his family. Knowing the scale of the upcoming disaster and the exceptional importance of human survival, we cannot eliminate the possibility of the existence of many other ships around the globe. As we know, before the first cosmonaut, Yuri Gagarin, there were many backup doubles who perished on duty. But people have a tendency to forget the ones who did not make it and only remember the names of the heroes who survived. Similarly, with Noah there could be other ships that just did not make it to shore. Christopher Columbus discovered America sailing on three ships, but not many of us know the names of the captains of the other two ships. Had Columbus perished at sea during his first voyage, the world would know the names of the other captains.

In order to answer many, many questions regarding the Great Deluge, we have to try to find our legendary Ark and here is some evidence of its existence.

Ancient Armenian historian Pavstos Buzand has stated that in the Fourth Century AD, Archbishop Hakob of Mtsbin organized a holy pilgrimage to Noah's Ark, but halfway there he was stopped by God, who appeared in his dream and told him that no mortal human being can see or touch the Ark.

In the hot summer of 1916, during the Russo-Turkish war while conducting air observations of the Turkish troops, Russian Imperial Air Force Lieutenant Roskovitsky noticed a half-frozen lake on the side of Mt. Ararat. As he flew closer he saw the semi-submerged hull of a large ship. The report was sent to Nicholas II, Tsar of Russia, who urgently organized a military expedition to the site. The urgency of the expedition is quite understandable because the region around Mt. Ararat in that year was experiencing unusual thawing. After weeks of painstaking ascent, the group eventually reached the site and found the Ark. A complete report with measurements, plans and photographs was sent to St. Petersburg. Unfortunately, the Tsar never received the reports due to the outburst of the Great Revolution of 1917 in Russia. Along with the report that vanished, the Tsar was murdered in the chaos and madness of the revolution.

Another detailed description of the Ark was done by a person who claimed that he had been on the Ark in his childhood with his uncle. His name is George Hagopian and he was interviewed by Rene Noorbergen (1928 – 1995) who later described it in his book *Secrets of the Lost Races* (1977). Here are portions from their interview, which I think are the most comprehensively descriptive of the shape and location of the Ark.

"I first went there when I was about ten years old. It must have been around 1902. My grandfather was the minister of the big Armenian Orthodox Church in Van, and he always told me stories about the Holy Ship and the Holy Mountain.

"And then one day my uncle said, 'George, I'm going to take you up to the Holy Ark,' and he packed his supplies on his donkey, took me with him, and together we started our trek toward Mount Ararat.

"It took us almost eight days from the time we left Van to the moment we got to the place on the Holy Mountain where both my grandfather and my uncle had said the holy ship had come to rest.

"I do remember that one side of the mountain is impossible to climb. My uncle and I went through Bayazit, close to the border, and climbed the

mountain from the direction of Azerbaidzhan. [The teller here most likely refers to the province of Azerbaidzhan, located on the North of Iran]

"It was resting on a steep ledge of bluish-green rock about 3,000 feet wide. Another thing I noticed was that I didn't see any nails at all. It seemed that the whole ship was made out of one piece of petrified wood. I could even see the grain of the wood, even though the ship had already turned to stone.

"I remember small holes running all the way from the front to the back. I don't know exactly how many, but there must have been at least fifty of them running down the middle, with small intervals in between...

"There were no windows in the ship, of that I am certain. And there was definitely no door in the side of the ship that I could see. No opening of any kind. There may have been one on the side I couldn't see, but that I don't know. That side was practically inaccessible. I could only see my side and part of the bow.

"The roof was flat, with the exception of that narrow, raised section that ran all the way from the bow to the stern and had all those holes in it. The sides were slanting outward to the top and the front was flat, too. I didn't see any real curves. It was unlike any other boat I have ever seen. It looked more like a flat-bottomed barge."

This intriguing and precious eye-witness story of the man who presumably has seen the legendary Ark was published by Rene Noorbergen back in 1977.

And again, instead of grasping one of the last strings connecting us with our past, our civilization did absolutely nothing to prove or disprove this precious information. The mechanism of reproduction and the following dedication to our offspring ingrained in our genes simply blinds our self-consciousness, where every single one of us is unique but all together we are just one mass led by emotions. I'm sure that for aliens, humans look like an ant colony where every member unconsciously performs his allegedly unique mission. We are creatures who constantly look into the future with certainty that one day future generations will find all the answers. Usually in case of water leakage, common sense tells us to look for its source. Yet why we search for the answers in the appeared pond is a paradox.

According to the above-mentioned witnesses, the Ark's presumed location would be at 3,900 – 4,500 metres above the present sea level. But as we saw it on the flooding simulation map of the Ararat Plain, the water level of the "Great Urmian Lake" could never exceed the 1,600 metres mark. There are only two possibilities for the Ark's "mystical" ascent to such a great altitude:

1. It was delivered there by the almighty waves of Earth's second reversal, where after the rapid melting of polar ice the planet's axis shifted again. (This scenario could be considered as a miracle because waves of such height would simply smash the ship into the slopes of the mountain).
2. During the following centuries, the Ark was elevated by the mountain itself.

In 1943 near the city of Uruapan in Mexico, in the cornfield of local farmer Dionisio Pulido, villagers reported hearing hissing sounds and the smell of rotten eggs. Mr. Pulido reported:

At 4 pm I left my wife to set fire to a pile of branches when I noticed that a crack, which was situated on one of the knolls of my farm, had opened... and I saw that it was a kind of fissure that had a depth of only half a meter. I set about to ignite the branches again when I felt a thunder, the trees trembled, and I turned to speak to Paula [his wife]; and it was then I saw how, in the hole, the ground swelled and raised 2 or 2.5 meters high, and a kind of smoke or fine dust-grey, like ashes-began to rise up in a portion of the crack that I had not previously seen... Immediately more smoke began to rise with a hiss or whistle, loud and continuous; and there was a smell of sulfur. (Wikipedia)

Within 24 hours, a volcanic eruption had generated a 50-metre-high scoria cone. Within a week, it had grown to a height of 100 metres. After periodic volcanic activity, the newly appeared mountain of Paricutin ceased in 1952 reaching its final height at 424 metres.

The birth and growth of the Paricutin volcano to a height of 424 metres in just nine years indicates that debris of the ejected satellites fallen to Earth have large areas of hot magma trapped beneath the fossilized crust, and Mt. Ararat is no exception. Its periodic eruptions could elevate its peak — and the Ark located on it — hundreds of metres at a time. Whether the Ark's gradual elevation was accidental or purposely stored for future generations, we do not know, but the fact of the existence of the "Great Urmian Lake" in the past and the occurrence of the Great Deluge prove its existence.

Our previous vision of the Great Deluge was nothing more than a myth. But now after considering our forefathers as real witnesses and not as some ancient lunatics, we can rediscover our entire existence. During the entire history of our postdiluvian civilization, we were underestimating the descriptions of all the strange events of antiquity and now it is time for revelation:

- Only the sudden dislocation of "Ice Age" poles to the equator could provoke such immense *"forty days and forty nights"* precipitation on Earth. That is why, being located in the basin of the former "Great Urmian Lake," *"The Ark went upon the face of the waters"*. That is why *"the mountains were covered"* and that is why *"the ark rested ... upon the mountains of Ararat,"* one of the most elevated highlands of the region.

Epilogue

As we know, after the Deluge Noah offered sacrifices to God. If we try to conceive this information from an alien point of view, we may notice some strange terminology. In our everyday lives we cannot consider the invitation of guests as sacrifice because the meaning of sacrifice indicates the offering of something to others without using it ourselves. The main assertion here is that God and the antediluvian people had quite different food rations. Antediluvian people were vegetarians because right after the deluge God suggested to earthlings to consume meat. If Noah's sacrifice was meat, then God was already a carnivore. Since gods came to our planet only in search of gold in order to save their own planet from losing its atmosphere, then they became meat-eaters way before they landed on our Earth.

God's postdiluvian permission to consume meat gives us an important clue that, besides the sudden shift in Earth's axis of rotation and following the catastrophic flood, there was at least one more drastic change, which lowered the volume of consumed proteins. We should not forget that the landscape surrounding the landed area of the Ark would look completely dead because, after 40 sunless flooding days, the entire flora and fauna would be devastated with little hope for a quick recovery.

And Noah the husbandman began, and planted a vineyard. And he drank of the wine, and was drunken; and he was uncovered within his tent. And Ham, the father of Canaan, saw the nakedness of his father, and told his two brethren without.

(Genesis 9:20)

Along with a detailed description of this global event and human survival, the ancient authors of the Old Testament describe the scene of a heavily drunken Noah. After the description of such a worldwide event we see a description of a drunken man. Quite strange, isn't it? But let us try to unravel that paradox.

If Noah ever had a weakness to alcohol, then according to his 600-year-old age it would not be his first drink, and his sons would not pay so much attention to their father's drunken condition. Authors of the Old Testament would not mention Noah's drunken condition if it were not odd to them, which means the people of the antediluvian world were not getting drunk, at least not so fast. As we know, the consumption of the same quantity of alcohol in different levels of oxygen in the air has a different impact on our body. Less oxygen in the air leads to faster dehydration and a greater effect of alcohol accordingly. A stronger influence of alcohol was caused by the drastic decrease of oxygen level in the atmosphere only.

What assertion we can make here is that during the Great Deluge, our planet lost a substantial quantity of oxygen. If before the First Great Catastrophe, thanks to high levels of oxygen in the atmosphere, all the living things had enormous sizes, then after the ejection of satellites when Earth almost burned all its oxygen, all living things shrunk to sizes appropriate for oxygen consumption. The same thing took place during the Second Great Catastrophe when the level of oxygen in the atmosphere went down even further. The air composition of the postdiluvian epoch had been drastically changed, and there were two major oxygen consumption factors:

1. Due to the simultaneous extermination of all organic matter, the decaying process consumed an unprecedentedly large volume of oxygen.
2. The simultaneous flooding of such vast areas of the earth led to the accelerated loss of oxygen due to being trapped during the formation of sedimentary rocks.

Scientific proof of a drastic change of oxygen levels in the atmosphere was found in a fossilized tree resin — amber. During its formation, along with trapping insects, it also traps bubbles of air. Analyses of the trapped gases show that Earth's atmosphere consisted of 35 percent oxygen compared to the present-day's 21 percent. And these amber samples most likely belong to the antediluvian period of our planet. If by any luck we ever find amber from the period preceding the ejection of satellites, then we could know exactly the level of oxygen in the atmosphere of Earth during the existence of gigantic dinosaurs.

This seemingly insignificant episode of Noah's drunken condition proves another of Zecharia Sitchin's very important proposal of the god's mission on our planet. As he stated, a race of gods came to our planet in search of gold in order to save their own planet from losing its atmosphere. If after each reversal, planets experience drastic drops in oxygen levels and weakening of magnetic field, then the gods' attempts to postpone the inevitable death of their planet is quite plausible and understandable. Considering the gods' departure from Earth, we can only guess about the fate of their planet and the fate of the gods themselves.

Unfortunately, everything in the universe has its beginning and its end, and imminent death awaits our planet also.

But what did our planet look like before the ejection of satellites? What we can surely assert here is that before the ejection of satellites:

- The earth's orbit was closer to the sun
- Due to the absence of satellites, Earth was spinning much faster
- It had a stronger magnetic field and better UV protection
- It was experiencing a major Ice Age on its poles and a stable moderate climate on 45° latitudes of both hemispheres
- Its atmosphere had a much higher level of oxygen
- Due to higher levels of oxygen, the decomposition of any organic matter was occurring at a much faster rate

After the ejection of satellites:

- The earth's orbit shifted away from the sun
- The acquisition of a satellite around Earth brought to a deceleration of rotation
- The Earth's magnetic field became weaker
- The end of severe Ice Ages (capable to make an indentation on Earth's surface)
- A drastic fall in oxygen levels

After the Great Deluge:

- A change of Earth's atmosphere (mainly a drop in oxygen levels)

As we can see here, the only difference between the antediluvian epoch and our own epoch is simply a drastic change in oxygen levels.

"Be fruitful and increase in number and fill the earth. The fear and dread of you will fall on all the beasts of the earth, and on all the birds in the sky, on every creature that moves along the ground, and on all the fish in the sea; they are given into your hands. Everything that lives and moves

about will be food for you. Just as I gave you the green plants, I now give
you everything." (Genesis 9:2)

The Great Deluge had a final knocking down impact on all the prehistoric giants. Forty days and forty nights of a harsh and sunless environment had crushed the last bearing foundation of the coldblooded dinosaurs. Any giant that miraculously survived would be destined to die due to a rapid decline of oxygen levels necessary to sustain their large bodies. Their sudden mass extinction meant the following predominance of smaller species.

God giving permission to consume meat indicates one thing only — the antediluvian people were vegetarians. The abundance of organic food in the antediluvian epoch had ruled out the necessity to consume meat. As we know from the Old Testament, Adam was feeding in the Garden of Eden, and there is no indication in any ancient text about Adam consuming meat.

Identically, with our automobile industry, vehicles with internal combustion engines were created by us to consume gasoline. If there were no fuel, obviously it would be absolutely useless to create internal combustion engines. Analogically, God created humans to sustain their living by consuming the product most available and abundant in nature. The following reconstruction of our daily ration was dictated by the drastic changes of our food supply. And believe it or not, besides the changes in our bodies, any change in our consumption (be it food, water or air) leads to the mutation of our societies also.

As a comparison, the crime rate in economically advanced countries is always lower than in economically struggling societies. Analogically, the abundance and accessibility of food in the antediluvian epoch would eliminate many reasons to fight with each other. We can only guess about the idealistic society of the pre-satellite ejection period where the abundance of food for any living creature on Earth looked like a fairy tale story about people living in harmony with Mother Nature.

As we know, every living creature on Earth becomes more aggressive during its mating season and when it protects its offspring. By comparing the duration of mating season of animals with the all-season mating abilities of humans, we can conclude that aggressiveness on the part of humans is present throughout the entire year. Here we should not forget that no animal cubs live with their parents for 20 to 30 years, and as long as the offspring live with their parents, protective aggressiveness of parents towards strangers will never diminish.

As we can see, only abundance makes all living things kinder. The postdiluvian shortage of necessary for consumption "energy" has transformed all the living things on Earth even further. The single change of oxygen levels in the postdiluvian epoch had transformed Noah's descendants into more aggressive creatures. That single change has brought us constant conflicts among ourselves. The difficulty in acquiring food and water had redirected us on a path of looting and using the labour of others. Unfortunately, that sinful desire will never fade away because our Earth will never be the same again.

But let us imagine a scenario strange for our times where people live in a completely different environment, where there is abundance of food, water and sex (unfortunately, those are the main physical desires of every living creature on this planet, including humans). There would be no need to fight anybody or to plan a strategy to seize someone else's possessions because abundance would eliminate any reason to do so. Now, by imagining this ideal society, we can surely assert that our prehistoric ancestors were not primitive creatures as we were told. They were simply living in a much more favourable environment and had more peaceful societies than we have now. That is the reason of such a drastic change of human society and rapid industrialization of our present world. Competition to possess new territories and their resources had triggered insidious cells in our brain. Unfortunately, along with the gradual diminishing of Earth's resources, our sinful desire to dominate will steadily grow and wars will always accompany our societies in the future.

I do not know if it was God's intention or not but unrecognizable changes in Earth's climate, the composition of the air, the magnetic field and a totally different landscape led to the emergence of religion. That's right, religion. Our forefathers were telling and teaching us to believe that all the events described in the Old Testament were real. The imaginative part of our brain was accepting it as truth, while the logical part of it was experiencing the discrepancy between ancient texts and the reality of our present world.

If, from the times of the first people on Earth there were no major changes in the environment surrounding us, then all the biblical narratives of our ancestors would not sound so science-fictional to us. Drastic changes of environment have led to the distortion and misinterpretation of many precious texts of antiquity, but in fact ancient texts describe events that occurred in the environments that are completely different from our present.

When we read ancient texts, we always have a feeling that the information we get comes from some third party, somebody who witnessed the entire history of mankind as a viewer. If, according to the Old Testament, the only survivor of the Great Deluge was Noah and his immediate family, then the entire story about the deluge and the Ark's survival was written by Noah himself or one of his family members. In our days, logbooks are always being written by the captains of vessels.

As we know, Noah passed away at the age of 950. The reality of our present epoch indicates the human average lifespan as no more than a hundred years. But what would be our concept about our age after arriving on a different planet? Would we try to convert Earth years into the calendar of the new planet? Common sense tells us that it would be complete nonsense to compare calendars of two different planets. An apple can only be compared to an apple. Therefore, the chroniclers of the Bible wrote exactly what they were told by their ancestors, and that is why Noah's total 950 years was divided to 600 antediluvian and 350 postdiluvian years. Multiple mentions of Noah's age in the Bible are not coincidental. All dates in the Old Testament that precede the Deluge belong to the antediluvian epoch. Events preceding and during the deluge were documented by Noah himself, who obviously was using an antediluvian calendar. The day when he opened the hatch of the Ark was logically the first day of the new era and the first day of the New Year, accordingly.

We cannot assume that the survivors of the Ark started their pilgrimage towards Mesopotamia right after disembarking. The best thing to do in their situation would be to follow the receding waters of the lake as the main source of food. Considering that the majority of all the living things on the surface of Earth were swept away, underwater life would boom due to the abundance of food. Just the fact that priesthood traditions survived among places neighbouring Mt. Ararat to periodically visit the Ark proves the constant inhabitancy of this region from the time of disembarkation.

Considering that the region of the "Great Urmian Lake" was the motherland of Noah, it is hard to guess the feelings of our great patriarch. Some people would choose to stay in their birthplace, and for others it would be too painful to see their motherland as an unrecognizable landscape. What is certain here is that descendants of Noah's family have spread to create the future Babylonian, Sumerian, Mesopotamian and all the ancient civilizations of the region. Although it would be almost impossible to sustain a civilized way of living, the possibility of people's survival in other regions of our planet is very probable. Survival of the legend of Atlantis, the ability of Procelens of Greece and the Indian tribes of South America to sustain the memory of the moonless Earth proves their autonomous survival of the Great Deluge.

If Noah is the forefather of all Mesopotamian civilizations, then during the constant wars in this region, people constantly exterminate each other. During the entire history of our civilization, borders of kingdoms were constantly fluctuating; the prosperity and expansion of one kingdom meant the decline and extermination of others. Much like boiling magma where every newly inflated bubble overlaps the others, its eventual burst leaves a round imprint on the surface until the appearance of other newly inflated bubbles will completely overshadow any visible trace of the bubble preceding it.

Let us consider one particular region on Earth with two neighbouring settlements. During the war between two mightiest kingdoms, these two settlements related to each other ended up being located on different sides of the border. After many years of existence in different environments, inhabitants of these formerly related settlements absorb all the traditions, culture and propaganda of the kingdom they belong to. Eventual conflict between kingdoms will lead to a battle between these two formerly friendly and related settlements. As we see on the lessons of our history, the relationship of people is getting easily overlapped by differences in ideology.

If we ask anyone on our planet "What is the most valuable thing for us?" the obvious answer will be "LIFE" and "HEALTH." We like to inflict pain on each other and to threaten to deprive someone of their right to live. If we look at the history of our civilization, we can see that all the wars break out due to differences in ideology, religion, geopolitical interests and the simple desire to acquire the wealth of others.

Any religion on Earth is teaching us to love and respect others, but for some reason we tend to idolize the preaching itself and completely forget about its content: *"Love your neighbour as yourself."* It always reminds me of the episode from Genesis when God asks Abraham to sacrifice his own son.

And Abraham stretched forth his hand, and took the knife to slay his son. And the angel of the LORD called unto him out of heaven, and said: 'Abraham, Abraham. Lay not thy hand upon the lad, neither do thou any thing unto him; for now I know that thou art a God-fearing man, seeing thou hast not withheld thy son, thine only son, from Me.' (Genesis 22:11)

It would be helpful to hear God's intonation and see God's face at that moment. Maybe it was expressing horror from the mankind he had created, where a mortal human was ready to slay another human (not to mention that it was his own son) just for faith's sake. Majestic "faith" in fact always overshadows our minds and forces us to forget about intellect and humanity. It would definitely grieve and disappoint God to see Abraham's readiness to kill his own son as well as it would grieve and disappoint any parent to see the fighting of his/her children.

Our constant wars on religious bases can be compared with a fight between people with amnesia who found themselves trapped on an island. While being unable to remember how they got to the island, people began to analyze possible variations. At first, they were sharing their assumptions of their survival. Gradually, seemingly peaceful debates began to turn into arguments with only one consensus that all of them were brought here by someone mightier than them. Whether because of loss of memory or the unwillingness to come to an agreement, everyone eventually began to push their own vision of creature(s) that could possibly have placed them on this paradisiacal island. In the end, a fight between them was inevitable and people with similar visions began to group together to eliminate the rest. And this seemingly funny paradox is called humans.

Chronological table of major events on Earth

- A powerful spark caused by close passage of a (substantial by size) celestial body.
- Formation of Earth and its gradual cooling.
- Gradual cooling leads to the planet's gradual acceleration of rotation.
- Gradual acceleration of rotation leads to a stronger magnetic field.
- Boiling waters of hot interior areas penetrate towards the planet's cooler exterior.
- The appearance of rapidly evaporating boiling ponds, lakes and seas lead to the creation of atmospheric circulation and appearance of atmosphere itself.
- Fresh water nourishing the highlands of the planet eventually led to the appearance of biological life.
- The constantly expanding atmosphere leads to the appearance of larger species of flora and fauna.
- Earth's rotational acceleration and absence of any satellites lead to the occurrence of Ice Ages.
- After the birth of the next member of our solar system, all the planets are shifting away from the Sun, causing lesser exposure to sunlight, weakening of the magnetic field and

inevitable loss of atmosphere. Earth is visited by a race of gods in search of gold to save their own planet from losing its atmosphere.

- The creation of humans to help the gods in their desperate need to mine gold. The race of giant people coexist with species of other giant animals (dinosaurs, mammoths, etc.).
- After the recurrent reversal of Earth, the planet turns into a position favorable for growth of one single water basin — the "Vampire Ocean."
- Gradual expansion of the "Vampire Ocean" leads to starvation and mass extinction of the majority of all living things who used to flourish at lower altitudes of the globe.
- Growing danger leads to the departure of the gods.
- The First Great Catastrophe. The birth of satellites and the shift of the planet away from the Sun. Mass drowning and suffocation of all living things located in lower altitudes. Drastic changes in Earth's magnetic field lead to the gradual mutation and shrinkage in size of all survived species of flora and fauna.
- The epoch of recovery, mutation and regeneration. Adaptation of the species that survived to the new environment with less intense sunlight and a weaker magnetic field. Plants, animals and degenerated people that miraculously survived gradually spread toward the remaining continents.
- The gods return and the appearance of early civilizations. The epoch of the second generation of humans living on Earth.
- Formation of a severe Ice Age due to recurrent reversal of the planet where both poles are dislocated towards dry land. Global drought.
- The Great Flood, survival of people and animals living on higher altitudes. Following mutation of all the living things on Earth, gradual appearance of modern looking humans. Gradual development of degenerated tribes and birth of our civilization.
- Gods' departure

Printed in Great Britain
by Amazon

60930134R00127